Bachelors of Science

In the series

Themes in the History of Philosophy,

edited by Edith Wyschogrod

Bachelors OF SCIENCE

Seventeenth-Century Identity, Then and Now

Naomi Zack

Temple University Press

Philadelphia

Temple University Press, Philadelphia 19122

Published 1996
Printed in the United States of America

∞ The paper used in this publication meets the requirements of the American National Standard for Information Sciences—Permanence of Paper for Printed Library Materials, ANSI Z39.48-1984

Text design by Erin Kirk New

Library of Congress Cataloging-in-Publication Data

Zack, Naomi, 1944–
Bachelors of science : seventeenth-century identity, then and now / Naomi Zack.
p. cm. — (Themes in the history of philosophy)
Includes bibliographical references and index.
ISBN 1-56639-435-X (cl.: alk. paper) .—
ISBN 1-56639-436-8 (pb.: alk. paper)
1. Philosophy, Modern—17th century. 2. Empiricism—History—17th century. 3. Philosophy and science—History—17th century. 4. Identity (Psychology) 5. Philosophers—History—17th century. 6. Scientists—History—17th century. 7. Bachelors—History—17th century. 8. Feminist theory. 9. Sex role—Europe—History—17th century. 10. Race awareness—Europe—History—17th century.
I. Title. II. Series.
B801.Z33 1996 95-52077

To my mother, Ruth Zack, 1914–1983

Discerne of the coming on of yeares,
and think not to do the same things still;
for age will not be defied.

—FRANCIS BACON

Contents

Preface

During my first year of graduate study at Columbia University in 1966, John Locke's *Essay Concerning Human Understanding* was required for the reading comprehension exam. This assignment was part of a generally ahistorical focus on the history of philosophy. Ideas from the past were expected to be imbibed, exposited, and then discussed critically as though someone had just come up with them. Although it was explained, in keeping with the tradition of classical scholarship, that ancient Greek philosophers thought with different concepts, this was presented mainly as a problem of translation. It was assumed that past philosophical thinking could be understood without understanding the culture in which it had first taken place.

My concentration on Locke was similar to earlier experiences with strange texts: in my junior year of high school, I read a section from Aristotle's *Nicomachean Ethics* for a class report in English; during my senior year in college, I read Martin Heidegger's *Being and Time* for a report in William Barrett's course on existentialism at New York University, and also struggled with Aristotle's *Metaphysics* for another course. My method in these assignments was to try to make sense of the texts "in their own terms" and I said that was what I was doing, at the beginning of my college papers. It did not occur to me that I was thereby avoiding a problem of translating concepts, or meanings, and my professors never commented on it. To understand a historical text "in its own terms," whether it is written in one's own language or translated into it, is to enter a limbo of meaning between the present and the past. For example, understanding Locke "in his own terms" is not understanding him in our terms, because our terms—that is, our meanings of the words he used—are different from his. Nor will Locke's text alone contain the key to understanding him in the way he would have been understood by his contemporaries.

As a student, I did not know that "foreign" concepts need to be understood on the basis of an understanding of other concepts, and events that were—or are, if the foreign writing is contemporary—present in the culture of original use. And I did not learn this by the time I received my Ph.D. in 1970, or during the next two decades when intellectual critical theory became incorporated into academic scholarship across the curriculum. I was absent from academia, and philosophy as an academic subject, during that revolution. When I returned to both in 1990, I saw that a change was in place through a new self-consciousness of the relevance of culture to thought, on even the most abstract levels.

When I began to teach full-time at the State University of New York at Albany in 1991, I had two lines of scholarly research. The first, racial theory, was partly a response to my own situation in academia. The second was a more mainstream study of John Locke's political and social philosophy, specifically his concept of ownership.

My work on race resulted in a book, *Race and Mixed Race* (Temple University Press, 1993), in which I analyzed American racial categories of black and white. I understood the main problem with these categories to be the lack of scientific foundation for any concept of race, and I was left with the realization that modern concepts of race derive from the eighteenth- and nineteenth-century pseudoscience that rationalized European colonialism and chattel slavery. It therefore seemed important that there was a period in European history when the idea of race as we know it now did not exist, and that the seventeenth century was the last time that Europeans had contact with non-Europeans without racializing them. Given that, I wondered what else about seventeenth century identity was radically different from later modern constructions. I already knew from my Locke research that ideas of ownership and rights were inextricably linked in the seventeenth century, and on the basis of contemporary concerns, questions about gender, class, and ecology came to mind. I could not pose those questions properly before acquiring a stronger historical sense of seventeenth-century culture and relating it to what I knew about philosophy of science and the history of science (I studied the former as a graduate student and have taught both in recent years). My partial answers to questions of how human identity was different in the seventeenth century from earlier and later constructions make up this book.

Acknowledgments

I wrote and revised the manuscript during a United University Professions Nuala Drescher McGann Affirmative Action leave award for 1994–95. I thank Bonnie Steinbock for supporting the project as a proposal for leave while she was chair of the Department of Philosophy here at the University at Albany. The award was augmented by travel funds to present a paper, "Sex and Race in the Seventeenth Century," at the American Philosophical Association in San Francisco in April 1995, and by funds for clerical assistance.

I have a scholarly debt to Robert G. Meyers, my senior colleague in the Philosophy Department, who has reinforced my sense of the importance of history to studies of the history of philosophy. He suggested leads for research, including the work of Barbara Shapiro and Maurice Mandelbaum, and we had several discussions about Richard Popkin's paradigm, as well as other aspects of my work in progress. I have also benefited from discussion with Linda Nicholson, Laurie Shrage, John T. Sanders, Wendy Donner, Berel Lang, Ron McClamrock, Ruth Sample, W. Bruce Johnston, and Ed D'Angelo. John Patrick Day was encouraging about my historical approach to John Locke's position on ecology in a paper I gave called "Locke and the Indians" at the Tenth International Social Philosophy Conference, at the University of Helsinki in August 1993. I also thank Thomas R. Martland for ongoing encouragement.

Temple University Press's four or five anonymous external reviewers of the book proposal and manuscript raised thought-provoking questions about my criticism of the feminist critique of early modernity, the method used in Part III, my general claims about identity, and several details of history and interpretation. I am grateful for their insights and conscientious attention.

Edith Wyschogrod, series editor of this volume, provided very constructive advice on evolving versions of the manuscript, and I appreciate her consistent support of the project. Doris Braendel, senior acquisitions editor at Temple University Press, was careful in tracking the varied critiques of the manuscript and I thank her for being so steady. David Updike copyedited the final manuscript with great care.

Once again, the final prepublication manuscript has benefited from Thomas Reynolds's meticulous clerical assistance.

Bachelors of Science

Introduction

Philosophy, History, and Criticism

A living being desires above all to *vent* its strength—life as such is will to power.

—FRIEDRICH NIETZSCHE, *Beyond Good and Evil*

Angle of Inquiry

There was so much change during the seventeenth century that we continually return there in order to understand the modern period. The disputes about the ideas of seventeenth century thinkers will never be resolved because our interpretations depend on our own changing cultural contexts. If we will never be done with the ideas of Descartes, Hobbes, Locke, Newton, and the like, neither can we escape the historical events of the period. The political, economic, and ecological events caused other events in chains that will always bind the heirs of Europe and its colonies. We cannot get rid of the past because it shapes the present. Although, of course, our understanding of historical events is as much a product of our culture as is our understanding of past ideas.

Moments in the history of philosophy, also, can only be understood through present philosophical (as well as cultural) prisms. My understanding of seventeenth-century political, scientific, and social ideas in this book is motivated by the present critical consensus that the Enlightenment left a bad legacy for women, nonwhites, and natural environments. The seventeenth century was not yet the Enlightenment, but most scholars see it as the conceptual and historical foundation for the events and ideas of the Enlightenment. My perspective begins from the assumption that the seventeenth century was not yet part of the modern period (although I follow the convention of referring to it as the early modern period), and I look at some of the seventeenth-century ideas of identity insofar as they are relevant to contemporary concerns. My considerations of historical and nonphilosphical cultural contexts is meant to go beyond ahistorical history of philosophy, but it is not directly opposed to traditional studies in the history of philosophy or to critiques of such studies, such as feminism.

Our contemporary academic consensus about the bad legacy of the Enlightenment is sometimes not carefully connected to the intellectual history of that legacy. Although we know well enough how to extrapolate the bad ideas behind the bad legacy, we do not pay enough attention to the connections among thinkers, their social situations, and their intellectual history. I therefore focus on some of the empiricist thinkers of the seventeenth century in order to deepen the understanding of what they thought from their own standpoints, rather than exclusively from the standpoints of those who have inherited and suffered from the bad legacy. My lever for this situated analysis is a "trope" of the English *bachelor of science,* as both an unmarried man and an individual whose life's work was centered on the New Science, and to a lesser degree, on the New Politics. These bachelors of science were inventing themselves as much as they were creating contexts for redefining women, the natural world, and those who came to be conceptualized as the inhabitants of the present Third World. They were, for the most part, younger sons, men without property and thereby men who were not themselves patriarchs, although their patrons surely were. They invented themselves as empirical thinkers in a New World Order of scientization, industrialization, exploration, and exploitation. I will argue that this self-invented social, political, economic, and intellectual type was contingently male, although perhaps necessarily European. Entrenched intellectual history has already connected the rise of empirical science with Protestant religion. Racially, however, the bachelor of science was not self-consciously white because modern concepts of white and nonwhite races had not yet been constructed. It is not even clear that modern concepts of the self were then in use. Therefore, as modern constructions of gender, race, and the self are falling apart in present scholarly criticism, it makes good sense to reexamine the seventeenth-century thinking that is conceptually prior to these constructions. This move also ensures against throwing out too many babies with the bathwater in the course of racial and gender critiques.

Inconveniently, the problem of preserving natural environments has no premodern resting place in the seventeenth century. In contrast to the undoing of race and gender, it is still not clear that there is currently a critical scholarly commitment, within any Western European academic discipline, to the undoing of artificialization. We can imagine androgynous, meritocratic, raceless cultures—as intellectuals, we partly inhabit them some of the time—but we cannot imagine a future without some next stage in technology. This could mean that human beings are unlikely ever again to be able to survive without artificialization, or that they are unlikely to choose to do so. The ongoing momentum of artificialization is accompanied by historical resistance to according rights or personhood to nonhuman life. This resistance may be due to greed, human needs, or the ease with which humans overwhelm and disregard nonhuman life.

At any rate, the seventeenth century yields few clues for emancipatory approaches to the problems of artificialization at the expense of nature. Nature was newly conceptualized as a subject of seventeenth-century science and technology, in ways that reflected a new divinity of men in that period. As a subject of science, nature was inert and quantifiable, but it was also evidence of God's design. As an object for technological manipulation, nature was simply God's gift to man. Women, children and non-Europeans were also viewed as exploitable material by European adult males. However, the de-animation of nature, or its relegation to physical "body" as opposed to nonphysical "mind," does not fully explain the acceleration of technological exploitation in the early modern period because women, children, and non-Europeans were acknowledged to have minds and souls, as well as life. If something new did happen to nature in the seventeenth century, it was as much a result of new ideas about those who manipulated it as it was the result of new ideas about nature itself. In that sense, the seventeenth-century nature of nature points back to to the seventeenth-century nature of man, that is, to questions of human identity.

The Subject and Its Limitations

The methodological subject of this book is identity. Identity is a contemporary critical concept that has varied meanings across emancipatory and liberatory projects. I mean to evoke standing intuitions about those meanings and broaden the concept to include the characteristics of a self that enable a particular type of self to act effectively in the world, from a given external situation. The external situation would be partly determined by cultural factors, such as politics, money, and social status, as well as more seemingly natural variables for identification by others, such as sex, race, and age. Identity is what enables a being that has it to change its situation and to choose and achieve goals. An individual without identity in an oppressive situation would probably be unable to resist or change the situation. Indeed, the current liberatory and emancipatory traditions posit identity as a tool for resisting and overcoming oppression. But when the concept is broadened to include agency, identity necessarily becomes more than the characteristics ascribed to those oppressed by their oppressors, as well as more than the characteristics that those oppressed recognize in their oppressors.

A supplementary motivating concept is needed to explain the inventions and changes in identity that took place during the seventeenth century. I think that motives to acquire and increase personal *power* can be assumed. Human beings do not aimlessly choose to be in some ways rather than others or struggle with how they already are, for no reason. Identity is worth having for the sake of increasing power. That is, if European males first invented new social, political, and intellectual attributes for themselves, exclusively, and denied identities to women, non-Europeans, and nonhuman beings, it was to increase their own power.

Within these terms of identity and power, practices of protest, revolution, resistance, resentment, and criticism also reconstruct and invent the identities of those engaging in them for the sake of increasing power. The concept of power is broad enough to include "utilities" of many kinds, such as survival, wealth, influence over other people and things, authority, happiness, and knowledge. For all of these utilities, it is better for an agent to maximize her own power, even if doing so leads to a minimization of the power of others—although it might be best if all identity types (and "tokens") were equal in power.[1] Thus, critical studies of race, gender, and environmentalism are empowering when they redefine or analyze away standing descriptions of women, nonwhites, and nature. The same holds for the identity and empowerment of children, which I consider to be an important subject in this book, although it slips out of the broad emancipatory picture because it is a relatively new subject there.

I draw on philosophy, history, biography, and intellectual history in order to investigate identity along the axes of agency and power. As a result of this interdisciplinary combination, there are omissions, imbalances, and a methodological untidiness that should be disclosed now, before individual chapters are surveyed.

First the omissions. As already noted, the treatment of ecological issues cannot be conclusive, due to both historical and cultural limitations. Art, poetry, fiction, theater, diaries, essays, and popular reading are not discussed, except for occasional brief references that pertain to intellectual history. Very little is said about law, economics, and the content of the natural sciences.

The imbalances in the work concern gender in contrast to race, and male in contrast to female gender. I write more about gender than race because the seventeenth-century bachelors of science were all European, which means they belonged to the group that was later designated as racially white.[2] The subjects are usually male because the bachelors of science were men.

Methodologically, the inclusion of biography and history is haphazard in comparison to specialized work in those fields. Traditional philosophers already coexist with historians of ideas, but both seem to have an unspoken aversion to history as historians and biographers reconstruct it. Intellectual biography has not yet been taken up by the emancipatory and liberatory traditions in the seventeenth-century cases in which it intersects with philosophy. There is therefore not much to consult for accepted ways of combining philosophy, the history of ideas, history, and biography from the standpoint of contemporary concerns about gender and race. If my experiment is fruitful, it should result in more systematic methodological standards.

Summary of the Book

Part I outlines the general epistemological context behind the more specific social, intellectual, religious, political, and professional characteristics of the seventeenth-century bachelors of science. In chapter 1, I begin with the contemporary feminist analysis of the new science of the seventeenth century because I share the goals of this analysis. However, I argue that aspects of the feminist critique, as developed thus far, have internal conceptual problems. Also, a feminist critique that proceeds from the standpoint of female gender as it was constructed after the seventeenth century would not be sufficiently contexualized to explain the motivations behind the development of seventeenth-century natural philosophy and philosophy of science. My criticism of the feminist critique in this chapter is meant to clear a path for an analysis of seventeenth-century empiricism that takes into account the intellectual history and intellectual biography of the philosophers in question.

Although René Descartes is not traditionally classified as an empiricist, he was committed to the practice of the new science. In chapter 2, I situate Descartes' skepticism in the historical context of the intellectual and religious-political problems he confronted. In that context, Descartes' famous doubt was not so much an expression of alienation from women and nature as it was a philosphical device meant to make science acceptable to the Church. Thus, Descartes' doubts were hyperskeptical only in order to highlight his positive program for scientific knowledge, which arrived at the end of a century-old tradition of skeptical argument in Reformation and Counter-Reformation theological debate about knowledge and epistemology. Although Descartes was friends with both Pierre Gassendi and Marin Mersenne, he rejected their *via media* answer to skepticism about scientific knowledge and unsuccessfully attempted to ground science on theology.

Chapter 3 applies Richard Popkin's contemporary intellectual-history scholarship on continental European skepticism to the *via media* aspect of English empiricism. Popkin's paradigm provides a bridge between the kind of continental empiricism that Descartes rejected and the English natural philosophy that did not initially aspire to Cartesian ideals of certainty. Robert Boyle revised the continental *via media* approach to science through his corpuscularian revision of Gassendi's atomism, and John Locke provided the philosophical foundation for that scientific revision. Although Francis Bacon gave the new science its ideological start as a pragmatic enterprise, his Platonic theoretical ideals fell outside of the English tradition, as did Thomas Hobbes' rationalist materialism.

In Part II I address the identity of the bachelors of science through analyses drawn from history, biography, intellectual history, social and political philosophy, and philosophy of science. Chapter 4 takes up the subject of their bachelorhood, beginning with the suggestion that recent demographic conclusions from social history block a modern psychological interpretation of their single status. For these bachelors, there was no analogue for later modern romantic motives for marriage. Few in their time married for love or even chose their own spouses; men and women married to gain and secure property and social status, as their parents directed. However, Descartes, Boyle, Hobbes, Locke, and Isaac Newton did not need to marry in order to succeed in their primary activities. Given that emotional needs may not have been foundational for the human self or person at that time, it may be that in the seventeenth century human identity was radically different from ways of being we take for granted now. This leads to the subject of formal, philosophical accounts of human identity.

Chapter 5 begins the philosophical analysis of seventeenth-century personal identity with Locke's theory. The chapter on personal identity in Locke's *Essay Concerning Human Understanding* is usually interpreted as a reduction of personal identity to memory. However, when we pay attention to his focus on *forensic* or legal persons and his "sameness" meaning of identity, Locke's theory supports the alternative reading that he is primarily interested in the sameness of a person on Judgement Day to the individual that person was in life. Still, Locke's preoccupation with conscience in this religious context ensures that the Lockean self is self-reflexively aware of itself; and its lack of a foundation in substance may be useful for present critiques of essentialism. More speculatively, the absence of strong, other-directed affect in the Lockean self may be promising for contemporary feminist critiques of traditional altruism and authoritarian family structures insofar as they are oppressive to women.

Chapter 6 continues the discussion of philosophical identity into seventeenth-century concepts of citizens and owners. The English Civil War idea of *propriety* merged absolute human rights with the rights of citizens to acquire and keep property. For Locke, the protection of ownership was less a moral matter than a natural consequence of his metaphysics of acquisition. The solitary nature of the Lockean citizen-owner is a reflection of personal isolation, in terms of affective alienation, but also in ways that might be relevant to contemporary solitude.

Chapter 7 is a bridge between my discussion of philosphical identity in chapters 5 and 6 on one side, and my discussion of professional identity in chapters 8 and 9 on the other. In chapters 8 and 9, I examine directly the kind of professional identity that developed in the context of the practice of the new science. The need for the institutional structure of the Royal Society is related to the lack of more traditional support for science on the model of Catholic Church support of learning on the continent. Seventeenth-century English Protestant factionalism is therefore the subject of chapter 7. English religion was imbedded in political disputes in ways that made religious toleration primarily a pragmatic strategy for keeping civic peace. Overall, there was less meeting of minds, even within the membership of the official Anglican Church, than historians of ideas have cared to admit. I offer the Locke-Stillingfleet dispute as an example of the parting of the ways between English Protestantism and empiricism. Nevertheless,

English religious institutions, while they did not positively support the new science, did not censure scientific innovation as the Catholic Church had censured the heliocentric theory and continued to exert pressure on the new mechanical philosophy.

Chapter 8 begins with a description of the continental antecedants to the Baconian pluralism of the early Royal Society. Through Boyle's philosophy of science, Hermeticism was officially removed from English science. Boyle developed the corpuscularian theory and propounded it for heuristic reasons, with an emphasis on the importance of actual experiments and limitations on scientific claims about reality. In keeping with Boyle's pragmatic and constructivist approach to science, the early fellows of the Royal Society cultivated intellectual virtues of tentativenes, cooperation and humility.

In chapter 9, I present Newton's work and personality as an antithesis to the ideals of earlier English empiricism. Newton, unlike Boyle, was a realist about scientific theory, and his solution to the hypothetical nature of his own theories was to insist that they were not hypotheses but descriptions of reality. Newton himself was competitive, secretive, and aggressively protective about his own work. These traits made it easier for scientific knowers to deify themselves, at the same time that they naturalized God as the ultimate craftsman and proprietor of the subject of scientific study.

In Part III I consider the roles and characteristics of children, nonwhites, women, and nature in the context of the new seventeenth-century scientific, commercial, and social establishment. In comparison to the identities of the bachelors of science, whether Baconian or Newtonian, these *others* were as yet unidentified: they had no positive identities that afforded them rights or agency, or any other increase in power.

Chapter 10 analyzes the most liberal early modern attitude toward childhood and early education as Locke developed it. Locke advocated kindness in child-rearing, in contrast to longstanding customs of (what would now be considered) abuse. However, his advocacy of motivation through shame and pride and his implicit assumption that gentlemen had a right to mold their sons to take their own places in society fell short of recognizing independent rights or agency for minors. The child was thereby a medium for middle-class social reproduction, rather than a person in its own right. To this day, liberal scholars continue to debate the appropriate degree of respect for the child as a person.

Chapter 11 addresses the most startling seventeenth-century difference in theories of human categories, the lack of modern female sexual and gender constructions. Although thoroughly subordinate to men in social status, women were conceptualized as sexually insatiable, inferior males with inverted male sex organs. There was little about motherhood or nurturance in their expected virtues. This absence of female sexual identity and later modern female gender characteristics, raises the intriguing question of whether sex ought to be held constant or gender ought to be held constant in transcultural investigations of male-female difference. It also renders remarkable the achievements of the first group of modern feminists that emerged in seventeenth-century England. Based on assumptions of intellectual equality between men and women, these early feminists questioned why women were subordinate to men in society, a question Hobbes had opened up, but only theoretically, when he wrote about "lord mothers" in the state of nature.

The question of how English dominance of the African slave trade coexisted with premodern rights rhetoric is the subject of chapter 12. Despite investment in the slave trade and general support of slavery on all levels of English society, there is no evidence of racial categorization in cultural or political thought at that time. Rather, trade and money were valued so dearly that actions that would have been criminal within the domain of English persons were not even subject to moral scrutiny when directed outward. However, the early rhetoric of the Royal Society did have racializing elements that exalted the particular intellectual "genius" of Englishmen.

In chapter 13 I compare the position of witches with that of magi and their empiricist heirs who denied any Hermetic association. As an intermediary between premodern authorities and the Devil, the seventeenth-century witch lacked identity of her own. This witch can be read as a passive point in a homosocial triangle that passed from her prosecutor, through her, to his fantasies about the Devil. Change in legal epistemology eventually led to a decline in the acceptance of evidence of witchcraft in particular cases. More broadly, empiricist knowledge requirements undermined general belief in the existence of witches, so that epistemologically, the bachelors of science were friendly to alleged female witches, while their Newtonian relaxation of empiricist standards for themselves meant that they preserved the tradition of the magi for themselves.

Chapter 14 affords the least degree of contrast between the seventeenth century and today. Nature now, as then, still lacks an identity apart from its intersection with human interests. The value of money received new theoretical and practical justification in the seventeenth century, and this accelerated conversion of natural resources to human wealth. Deep ecology would have been unthinkable then, and conservation was purely instrumental. Today, the effectiveness of deep ecology seems to require radical ecology to provide nature with the agency necessary for independent identity, at best a kind of metaphoric human power supported by enforced laws.

Part One

The Intellectual Context of the New Science

One

Feminist Criticism

> The advantage man enjoys, which makes itself felt from his childhood, is that his vocation as a human being in no way runs counter to his destiny as a male.
>
> —SIMONE DE BEAUVOIR, *The Second Sex*

Contemporary feminist criticism of the genesis of seventeenth-century empiricism zeros in on a bias against women and a destructiveness toward nature that appear to be foundational for subsequent philosophy and science. This kind of criticism, as developed by Carolyn Merchant, Susan Bordo, Evelyn Fox Keller, Jane Flax, and others, combines psychoanalytic concepts and object relations theory with a violated cultural metaphor of motherhood. Although the psychological and metaphoric emphases vary in complex ways among different writers, the general project may be ahistorical in ways that call for more intellectual history to understand the identities of seventeenth-century empiricists. Both the compelling nature of the feminist critique and its conceptual and contextual problems are particularly evident in the treatment of Descartes.

The feminist critique holds Descartes responsible for situating intellectual identity in masculine minds, as opposed to masculine *and* feminine minds, or masculine and feminine minds *and* bodies. This view of his role is based on the following interpretation of *Meditations on First Philosophy*. Descartes begins by doubting that he knows anything and in the throes of this disorienting anxiety, he posits himself as a thinking thing (*Meditations* I).[1] But Descartes is unable to include his body in his identity because knowledge of his body comes to him through his senses and imagination, which can be deceptive. He progresses from the indubitability of his own existence as a thinking thing, which is given by the very act of doubting, to his knowledge of the primary qualities (i.e., extension, figure, quantity, number, place and time) of material bodies (*Meditations* II).[2] This knowledge, which can be scientifically expressed in mathematical form, is secured only because God is good and not a deceiver. Both the existence of God and His goodness are given through an idea that is a representation of Him. The reality in the idea of God must also be present in its cause, so God is real, and if He is real, He must be good (*Meditations* III).[3] Overall, Descartes is not disturbed by the nonmaterial quality of the universe he derives from thought; on the contrary, he devotes himself to the formulation of rules that will keep his mind "pure" of the illusionary ideas that come through the senses, that is, through the body.

This interpretation of the *Meditations* does not contradict traditional philosophical exposition, although it is an oversimplication. For instance, Henry Frankfurt offers an interpretation of the *Meditations* as a heuristic exercise to lead philosophical neophytes to the distinction between scientific knowledge and ordinary belief;[4] and Amélie Oksenberg Rorty argues that there is ample textual evidence that Descartes fully appreciated the role of uncertain sensory knowledge for the content of science, if not for its first principles.[5]

But even where the feminist reading is restricted to Descartes' specifically rationalist quest for first principles that can be known with certainty, an added layer of psychoanalytic interpretation results in characterizing the *Meditations* as a *flight* from all things female. In Descartes' life, this flight is read as a *reaction formation* resulting from the loss of his mother when he was thirteen months old. The reaction formation attributed to Descartes as an individual is offered by some feminist scholars as an expression of a widely held combination of cultural beliefs during the early modern period. On this "psycho-cultural" feminist interpretation, industrial changes during the seventeenth century generally alienated men from nature and the earth.[6] Such alienation is read in contrast to beliefs about nature in the medieval period, when the earth was conceptualized as female—literally as a nurturing woman whose inhabitants were harmoniously part of her because she was their mother. That medieval organic relationship was ruptured by early mercantile capitalism when great forests were destroyed for shipbuilding, the earth was torn up by mining for metallurgy, and wetlands were drained and commons enclosed for more systematic agriculture. At the same time, the dispossessed rural poor had to migrate into towns and cities where they became day laborers and wage earners.

As a result of these economically driven changes in ecology, the earth was no longer seen as a cherished support of human life, but as a mere material thing—in contrast to the Cartesian mental thing—that could be observed, experimented upon, and divided into quantifiable parts with no guiding principles beyond the expedience of scientists, businessmen, owners, and craftsmen. Since all of the exploitable and manageable aggregate of nature was not only physical but female, the inquisitive, controlling, and aggressive male mind set itself off in dominating opposition to the passive, controlled female body.[7] This *cultural* female body included nature, the bodies of women, and the bodies attached to male minds as well.

The Cartesian mind-body and male-female bifurcation has also been applied by femininists to early English empiricism, through straightforward readings of the misogynistic language in Francis Bacon's presentations of the new science. Although Bacon did not demand the same certainty for the foundations of knowledge that Descartes did, he was skeptical about earlier philosophical systems and called for a "Great Instauration" for the practical benefit of mankind. Bacon laid down the inductive rules for observation and experiment, according to which, in a decade or so, he thought, all of the natural sciences could be fully developed. Nature was to be the object of this endeavor, and although "she" had to be obeyed in order to be commanded, Bacon indicated through sexual metaphor that this would be a masculine task of aggressively dividing natural philosophy into two parts—"the mine and the furnace." In a letter to King James I, Bacon compared natural philosophy to witch trials: "Neither ought a man to make scruple of entering and penetrating into these holes and corners, when the inquisition of truth is his whole object."[8]

Running parallel to this feminist psychoanalytic criticism of Cartesian rationalism and the focus on Bacon's sexualized empiricism are insights from object relations theory, which take this form. On a socially dynamic model, male and female gender identity are constructed through a *reproduction of mothering* in each generation. Children of both sexes have female nurturers in early childhood. Boys grow toward ideals of autonomy with cognitive and physical achievement dominant over affective ties so that they develop their male gender identity by separating from their mothers and becoming unlike them. Girls grow toward ideals of dependence with affective ties dominant over cognitive and physical achievement, so that they develop their female gender identity by remaining close to their mothers and becoming like them, to the point of becoming mothers themselves and reproducing this process of gender construction.[9] Thus, the male distancing from mother nature and the construction of intellectual identity and scientific knowledge in ways opposed to nature during the seventeenth century can be viewed as a cultural macro-mirroring of the constructions of male and female gender over generations in families. The feminist psychological claim is not that modern science was caused by the individual psychological pathology of its founders, but rather that the alienation from the earth and women in the world view of modern science is a universalization of male infantile experience.[10] And furthermore, it is claimed that this universalization resulted in a masculinization of the ideal practitioner of science, along ideas of alienated male gender, that is still evident in science today, especially in the (so-called) hard natural sciences.[11]

The feminist critique of seventeenth-century philosophy and science has added rich cultural context to traditional accounts of the development of Western knowledge. Nonetheless, this critique has built in conceptual problems that block further contextual understanding of seventeenth-century ideas. The application of psychoanalytic and object relations theory concepts to cultural and intellectual history, as well as philosophy, raises both empirical and abstract problems. The use of both types of psychological concepts together may be theoretically confusing. The mother earth metaphor, because of its poetic fragility, is easily overextended. A general claim that reason was masculinized is difficult to substantiate, although there was certainly a rhetorical masculinization of reason, and women were excluded from the projects of reason. Finally, even if the parts of the feminist critique presented above were right about the causes of seventeenth-century philosophy and science, critiques of causes of ideas cannot do the work of critiques of ideas. Nonetheless, the problems with the feminist critique of seventeenth-century philosophy deserve careful attention because the motivating feminist concerns about the position of women in the seventeenth-century intellectual climate, and subsequent developments from that position, are so important.

Problems with the Psychological Interpretation

Feminist criticism of early modern science sometimes proceeds psychoanalytically by hypothesizing strong affective motives that Descartes, for example, as a generic representative of childhood masculine gender formation in the early modern period, might not have recognized as his own. This is partly a problem with attributing early childhood psychological experience to an adult who shows no conscious signs of being motivated by it—that is, a problem with Freudian analysis. However, feminist criticism is not wholeheartedly Freudian. So it should be feasible to take a closer look at a feminist use of reaction formation, in isolation from the more theoretical labyrinth of full-blown Freudianism. (I reserve for later chapters the more general question of the applicability of any psychological model that posits deep emotional drives in seventeenth century individuals. It is assumed, for example, that men first loved the earth because "she" was experienced by them as a mother. But we do not know that men loved their mothers in the medieval or early modern periods, or that their masculine gendered affect would fit any current psychological norms based on the "health" of such love.)

How, then, can reaction formation be used to explain why Descartes and his contemporaries turned against both women and mother earth, and why the entire modern scientific and philosophical project was hostile to women and nature? We could assume that love for mother and nature was so strong that separation from them was dealt with by taking control over the separation itself, in a cold and hostile way. This reaction formation of aggressive separation and control must have continued through the ongoing scientific and industrial revolution, if reaction formation was a principal cause of that revolution or the revolution an expression of reaction formation. But in order for the revolution to continue as an effect or expression of reaction formation, there would have to be cultural evidence of the love of nature by those who were destructive toward it, and this evidence is lacking.

There lurks here a more general problem with the application of reaction formation to cultural events. It is not clear how a psychological theory about the behavior of individuals can be applied to large-scale cultural events. Even if it could be shown, for example, that due to his mother's death Descartes had problems in constructing an identity for himself that included his body, everyone influenced by his epistemology did not lose his mother in infancy. A feminist critic might claim that many people had experiences analogous to the death of their mothers, as a result of the destruction of the natural environment. But isn't the destruction of nature meant to be interpreted as a reaction? What is the separation from nature that causes the destructive reaction to it, insofar as the separation is read from the destruction? The only people who were separated from natural environments against their will were the rural poor who had to leave the land, but these were not the individuals who embarked on philosophical quests for certainty or programs of scientific experiment.

Object relations theory seems to lose its empirical base outside the context of a traditional, twentieth-century, middle-class nuclear family. If male gender identity is the result of separation from mothers, then in a culture where children of both sexes were separated from their mothers the structures of separation and identity should be the same, and the gender differences in question could not be explained by object relations theory. During the seventeenth century, in middle- and upper-class families, female as well as male infants were usually sent away from their mothers, shortly after birth, to be nursed by other women, in separate households, until they had survived infancy and could be weaned.[12] Therefore, seventeenth-century gender differences cannot be explained on Nancy Chodorow's "reproduction of mothering" model because both genders were forcibly deprived of mothering. In addition, the measure of masculine autonomy, as part of masculine identity, by patterns of affiliation would be thrown off by the presence of binding relationships among men. Virtually all religious, social, economic, intellectual, and political groups were exclusively male in the seventeenth century.[13] Since these groups were the media for fundamental cultural change, it seems highly probable that they were the site of important friendships and professional relationships.

Object relations theorists themselves relate that, in contemporary children's games, boys play in larger groups, for longer periods of time, with greater group stability over time, and more planning and discussion, than girls do, all of which could suggest that boys are more affiliative than girls.[14] This description of boys' play could support a claim that male identity is based on male-male relationships, whereas female identity is based on female-female relationships. In that case there would be no initial gender imbalance to reproduce through female mothering and same-sex affiliation, but merely a symmetrical gender difference in development. Object relations theory alone would thus be unable to account for the initial gender difference, although there would be room for theories about the segregation of children based on sex, as well as different cultural expressions of that segregation.

The application of psychoanalytic concepts usually proceeds on an assumption of inherent psychological drives, or at least strong emotions, in all men, everywhere, all the time. That is, although the psychoanalytic id is purported to be gender-neutral, feminists have read it as a transcultural model for male gender, insofar as male gender is their subject of criticism.[15] If such psychoanalytic concepts are applied to gender identity, then object relations theory cannot also be applied universally because that theory is based on the assumption that gender is the result of social activity (and roles) in human culture. If gender identity were universal and given from birth in a psychoanalytic sense, there would be no need for cultural explanations of its formation.

This tension between psychoanalytic concepts and object relations theory increases as wider empirical possibilities are considered. Suppose, for example, that a female child is raised by her father and identifies with his patterns of affiliation, or that a young male reacts to the loss of his mother by rejecting a masculine identity based on separation from her. How could one combine psychoanalytic concepts and object relations theory to interpret such cases? In clinical practice rules of thumb would probably be used and different theories and their hypotheses and exceptions could be tried, without excessive concern about abstract consistency. In cultural analyses as well, psychoanalytic concepts and object relations theory could be combined in narrative or literary criticism so that the culture is viewed as the individual writ large, nature and nurture interplay, and the tension between the individual and her culture determines the course of the individual's life, the individual being always "of her times." However, as applied to the early modern period, the aspects of feminist criticism in question seem to promise a model of analysis that need not rely on such creative improvisation.

If the values of feminism include the well-being of individual women, there might also be a conflict between those values, in particular cases, and critical evaluations of male behavior based on object relations theory. Indeed, some object relations theorists describe male gender formation as a process oppressive to women, and use this to criticize male gender and suggest how it can be revised into more nurturing and caring modes.[16] Within a feminist value system, however, the separation of male children from mothers could be viewed as an overall gain if it disrupts usual patterns of male gender formation that are oppressive to women. If male children are involuntarily separated from their mothers, then presumably they have no need to individuate themselves, as males, by hostile self-distancing from their mothers and other female "objects." The object relations theorist might here object that such males would be even more strongly gendered for not having had the usual contact with women early in life. But then, the concept of male separation from mother as a way through which men choose their masculine gendered identity loses its explanatory force.

There is also the question of psychological well-being, which almost all psychological theorists associate with good and sufficient mothering. In general, psychological accounts of early male separation from mothers tend to tie this separation to later failures in relationships with women, or to adult avoidance of women. This was presumably the case with Descartes, who never married and described the death of his former mistress as God's gift to prevent him from renewing a "dangerous commitment."[17] But what harm does this failure do from a feminist point of view since such early-separated males are less likely to come into close contact with women and thereby oppress them? The rejoinder could be that, in the case of Descartes, at least, his *ideas* were oppressive to women. This, however, brings us full circle because the Cartesian ideas were initially to be explained in psychological terms, but now the misogynistic psychological explanation itself depends on a misogynistic reading of the ideas.

Suppose we grant that Descartes and many other seventeenth-century thinkers were psychologically "unsound" and that their individual pathologies were the result or expression of wider cultural conditions. The concept of psychological pathology is itself highly sensitive to cultural contexts and care needs be taken that it not be used to carry an unstated ad hominem argument. Psychoanalytic theory has a history of providing new descriptions and explanations whose initial effect is insult and character assassination within the cultural context in which they are introduced. When the audience has, through repetition or historical changes in its own standards, become desensitized to what was initially seem as shocking and derogatory, the initial application of the theory can be read as a neutral description. At that point, the theory has to be applied in a more extreme way and the "patient" described in ways suggesting greater pathology for a reaction of shock and its attendant devaluation of the patient to occur. Two layers of psychological scholarship over Descartes' three famous dreams provide an example of how the devaluing aspect of psychological historical interpretation has shifted in the service of ad hominum.

In 1969, when Ben-Ami Scharfstein wrote about Descartes' dreams of 1619, he offered an eclectic but mainly traditional Freudian analysis of Descartes that emphasized his feelings of moral and physical inadequacy and his isolation, curiosity, and depersonalization. Scharfstein presented Descartes, not unsympathetically, as a lonely, courageous individual whose desire for firsthand knowledge gave his philosophical writings an enduring charisma.[18] This is how Sharfstein concluded his analysis:

> [Descartes'] analysis has come to be centered in the way in which he dramatizes his predicament of doubt and certainty and, forgetting precedent, begins from himself. As a beginning, this is permanent and it remains in the endless present of a lyric or autobiography. It is his need to be original and

> find his way, for himself, not in the company of parents or traditions, that we feel most keenly. . . . It is his dreaming that lives, or, rather, his dreaming concealed behind his mask of reason, or perhaps, his dreaming mingled insensibly with his open-eyed reason.[19]

In 1987, when Susan Bordo approached Descartes from a feminist psychoanalytic standpoint, she began by characterizing his dreams as "nightmares," thus:

> On November 10, 1619, Descartes had a series of dreams—bizarre, richly image-laden sequences manifestly full of anxiety and dread. He interpreted these dreams—which most readers would surely regard as nightmares—as revealing to him that mathematics is the key to understanding the universe.[20]

Bordo does not mention that the tradition of Cartesian scholarship has always called them "dreams," nor does she justify her departure from this tradition. Throughout her analysis of the *Meditations,* Bordo leaves the label "nightmare" in place so that it is Descartes' nightmares, and not his dreams, that are being shown to underly his philosophy. The reader should judge for herself how disturbing Descartes' dreams actually were. (I would not classify them as nightmares because they do not seem to hit the note of terror associated with nightmares, in my own experience. I am not sure, given the intellectual interest Descartes himself took in them, that they were even "bad dreams.") Characterizing philosophical inquiries as the expression of nightmares is a new interpretation that reevaluates them in a way that is shocking and intellectually discrediting. Nightmares are symptoms of psychological disturbance, and a philosophical insight based on a nightmare would be suspect on those grounds alone. But if Bordo's reader tries to search behind the effects of the nightmare rhetoric, no light has been shed on the grounds or norms of psychological health against which what were previously dreams should now be understood to be nightmares.

Overextension of the Mother Earth Metaphor

Mothers do nurture their young, and most cultures have recognized that they are thereby worthy of deference, protection, and respect. Some cultures have characterized the entire planet as a mother to human beings. The feminist criticism that locates the end of the cultural use of this metaphor between the medieval and modern periods, draws on an intuitive sense that it was morally wrong for scientists and early capitalists to turn on the natural environment and its creatures as a source of selfish gain. The violent acts committed against mother earth by men were, by implication through the metaphor, as wrong as matricide. However, the comparison of the planet to a human mother overextends the model of the mother-child relationship.

John Locke's argument that political relationships are not the same as paternal family relationships is directly relevant to this overextension. Locke's main task in the *Two Treatises of Civil Government* was to clarify the ways in which political relationships could be justified. He insisted that politics was a separate area of activity, distinct from both religion and family life. Against Robert Filmer, Locke showed how the patriarchal model in particular and the parental model in general failed to justify an unlimited and divine power of kings. Locke very crisply pointed out that parents have a duty to provide for their children that is completely discharged when the child becomes an adult. And while grown children owe their parents honor and respect, they are not obligated to obey them as adults, or to pay them back in kind for benefits received. That is, the parent-child relationship of nurture and obedience is temporary. By contrast, the relationship between sovereign and citizen is lifelong and is based, in principle, on the consent and benefit of the citizen from its inception onward.[21]

This same Lockean argument works against a maternal model for the planet. The nurturing of children by their mothers physically ends with the maturity of the child, while the physical dependence of human beings on their planet is lifelong. The kind of deference, obedience, and respect appropriate during the intense period of the child's maternal dependence does not automatically extend to ecological relationships. There are no grounds for extending the emotional values of the mother-child relationship to ecological relationships because those emotional values arise from relationships between particular human beings during restricted periods of time; if mothering has a purpose, it is to support the child until mothering is no longer necessary.

Unless the mother earth metaphor is accepted uncritically, it does not provide a foundation on which to criticize seventeenth-century changes in human relationships with the natural environment, or to criticize how conceptualizations of the natural environment changed. If the mother earth metaphor was abandoned by seventeenth-century scientists and capitalists, they cannot be faulted simply for that or criticized for what would have been a violation of the metaphor had they continued to use it. While the metaphor may provide gratifying imagery across a variety of cultures, it lacks the intellectual power attributed to it by some feminists.

The De Facto Masculinization of Reason

While women were excluded from intellectual and scientific work in the sixteenth and seventeenth centuries (and beyond), it was not yet the case that they were excluded because male intellectuals officially maintained that women lacked the required mental capacity to undertake such work. Throughout Europe at that time, women were excluded from all employment in which they would compete with men. Some writers have connected this situation with the widespread legal and political assumption, in a time of job scarcity, that the basic social unit was a male head of household. It was considered natural and socially appropriate for adult women to be married; and married women could not own property or be party to contracts independently of their husbands. Single women and widows, although they were often viewed with suspicion, had more legal and economic freedom than wives, especially when they had independent sources of income or owned property.[22] This begins to suggest how subordinate gender was tied to marital status and class position, although paternal pressures to marry for financial advantage to the family, and spousal pressures to bear children, were more intense for upper-class women.

Among the upper classes in both Protestant and Catholic countries, and within Catholic clerical cultures on the continent, there were a few female intellectuals in their own right, as well as women who corresponded with philosophers and scientists and who studied science, philosophy, and religion. None of the Catholic or Protestant sects of the period disputed that women had souls, could reason, or were educable. The writing of Teresa of Avila suggests that the late medieval identification of the mind with the soul as a thinking substance would have precluded any claim that women had souls but not reason.[23] Since Descartes also identified mind with soul, if women had souls on the Cartesian model, and they would have had to according to Church doctrine, then they had minds and, by extension, the capacity to reason.

Genevieve Lloyd points out that there was no basis in principle for Descartes to exclude women from the practice of his philosophical method. In fact, he had extensive intellectual interactions with women: he corresponded closely with Princess Elizabeth about philosophy; and he reluctantly, and fatally, went to "the land of the bears" to instruct the Queen of Sweden in philosophy.[24] Indeed, Descartes begins the *Discourse on the Method* with a claim that reason is equal in all human beings and that our opinions differ only because our thoughts "proceed along different paths" and we are therefore not thinking about the same things. Our equally good mental powers do not serve us all equally because we do not employ them "rightly."[25] Although Descartes uses masculine pronouns in this context, he obviously thought his methods could be used by women because Elizabeth lamented to him that her domestic duties interfered with her use of them. Lloyd also notes that Descartes told a correspondent that he intended his method to be accessible to women and that he wrote the *Discourse* in the vernacular so that it could be read by women who were educated at home and not in the schools that taught Latin.[26]

The final philosophical point about Descartes and the masculinization of reason is that if he is taken seriously in identifying himself with a "thinking thing" that is not a body—an interpretation not all Cartesian scholars agree with[27]—then he is something sexually neutral because his biological maleness would derive from physical characteristics. This means that in the absence of gender constructions that precluded it, there was nothing in Descartes' philosophical identity that excluded a woman from having that same kind of philosophical identity.

In contrast to Descartes' relative feminism in these matters, Francis Bacon explicitly called for a masculine science, and he considered the female influence over male intellectual activity on the continent to be enervating and contemptible.[28] Along the same lines, women were not usually allowed in the audience of the Royal Society during learned demonstrations and the group did not even have a female member until 1945.[29]

However, a case can be made that Bacon deliberately used sexual imagery as a rhetorical device to recruit adherents to the new science. Such unscrupulousness is not inconsistent with generally accepted accounts of Bacon's character. Even his most polite traditional biographers acknowledge that he was disloyal to both colleagues and patrons and not above scheming or taking bribes when in high office[30] (although exonerations of Bacon can be found in the relatively polite genre of nineteenth century historical conspiracy).[31]

James Stevens emphasizes Bacon's long-term project of developing persuasive ploys for winning over his audience. Bacon had a low opinion of the common reader but also realized that even philosophers, scientists, and kings were greedy, passionate, and self-important. Stevens points out that

Bacon was aware of the use of metaphor, myth, perceptual imagery, and the force of memory, as well as the effectiveness of motivation through vice. Between 1597 and 1625, Bacon worked on the relation of the content of science to its style of delivery before the public. In *The Masculine Birth of Time* he discussed looking for "quiet entry" into the minds of his readers so that philosophy could "select her followers through their appetites." In *The Advancement of Learning* he referred to a popular style that would create images and ideas to be stored in the memory and reproduced for future use.[32]

It would not have been beneath Bacon to use sexual imagery deliberately to popularize science by appealing to what he thought were the basest and most insistent impulses of his (male) readers. The absence of any attempt to censor this rhetoric and Bacon's huge success as a propagandist for English science underscore widespread negative attitudes toward women, in the seventeenth century and beyond. A reinforcement of these degrading attitudes can also be read into Bacon's texts. (Like the use of sexual imagery in contemporary media and print advertising, however, this does not in itself make the product advertised, or its functions, exclusively masculine or oppressive toward women.) This reading of Bacon's use of sexual imagery restores a level of sophistication to his writing that allows the reader to distinguish between chosen style and (sincere) self-hypnotism by metaphor. To believe that Bacon could not have presented his case for science without sexual imagery, in all of his writings, is to miss a cleverness in seventeenth century writing, generally, which could also dupe the contemporary reader into overlooking the nature of the self-interest and quest for power behind seemingly quaint prose.

Locke offers direct, though weak, indications that the second wave of English adherents of the new science did not intend to exclude women as a matter of principle. Locke wrote a female correspondent that there need be no significant difference in the way boys and girls were educated.[33] He also had intellectual friendships with women, which seem to have been more personal than Descartes'. For example, he found final comfort in the home of Lady Masham, née Damaris Cudworth, whom he may have almost married earlier and who became his principle biographer after his death.[34]

Upper-class women were recognized as part of the educated reading public for popular scientific works that presented Descartes' cosmology and "Newtonianism for the Ladies." In 1704, *The Ladies Diary,* an almanac containing scientific information, was first published.[35] While parts of Locke's and Descartes' writings and other early feminist inclinations had minimal practical effect against the widespread exclusion of women from intellectual and civic life, they nevertheless highlight the difference between the seventeenth-century attitude toward female capabilities and eighteenth- and nineteenth-century constructions of female gender that posited women as "naturally" deficient in reasoning ability. It therefore seems a mistake to assume, as feminist scholars sometimes do, that these later constructions of female gender were directly grounded in the constructions of rationalism and empiricism in the seventeenth century. Even though female intellectuals were more likely to be patrons of male intellectuals than to have patrons of their own, or any sort of public recognition for their intellectual work and study, there were female philosophers and scientists in the seventeenth century. Margaret Cavendish, duchess of New Castle, wrote seven books of philosophy and science that combined her own ideas of matter, motion, and epistemology with those of Locke and Descartes. Maria Sibyella Merian, in Germany, studied insects and plants. Astronomical data was frequently recorded in private homes by the female relations of male astronomers. When Elizabeth Cellier, an experienced midwife, petitioned James II for admittance to university medical training and Maria Winkleman, an accomplished and published astonomer, requested an appointment by the Academy of Sciences in Berlin, they were both denied, not because their competence was in doubt but because they were women.[36] This connection between sex and professional status, without the mediation of gender, had the same employment results as did later beliefs that women were incapable of intellectual work. But, and this is the point worth taking, the ideas of Descartes, Bacon, and their contemporaries were not yet sufficiently engendered to "masculinize" either philosophy or science.

To conclude this discussion, the connection of the position of women with the new ideas of Descartes and Bacon is no stronger than what is suggested in the quote from Simone de Beauvoir at the beginning of this chapter: The vocation (practice and contents) of philosophy and science did not conflict with male destiny and identity in the seventeenth century, but due to widespread social conditions, it did conflict with female destiny though not necessarily with female identity. As an addendum to this, it should be noted that the high value placed on pure thought by Descartes and his followers excluded not only women from philosophy and science, due to their other obligations, but also the majority of men, because their economic position in society required that they do physical work.

In the aspects discussed, the feminist critique overstates its case about seventeenth-century male philosophers by not going far enough into the historical conditions and the ideas of the time. From the standpoint of the history of philosophy, the most serious problem with some feminist criticism of Descartes and Bacon is that it dislodges them from their intellectual contexts and distorts their influence over their own and subsequent cultures. The failure of feminist criticism to show how the distancing of male thinkers from women and nature was an inherent aspect of seventeenth-century empiricism redirects the search for motives toward the intellectual context of the time. Feminist criticism, as developed thus far, does not tell us how Descartes, Bacon, or anyone else in the period approached their philosophical work from their own backgrounds and positions in society. We need to know what problems these thinkers and their contemporaries thought they were solving. That would be the first step toward turning the philosophical content back to the life circumstances of its authors, in order to relate their work to their identities. The new scientific identities they invented for themselves increased their power in ways such that, by comparison, women and other beings, who had no positive identities, were passive. That absence of identity will be my concern in Part III. In the next two chapters, beginning with Descartes, I relate the particular intellectual projects of the bachelors of science to the problems posed by pyrrhonic skepticism.

Two

Descartes' Doubt and Pyrrhonic Skepticism

> Experience does indeed show that such reputed sources of knowledge as memory of perception or testimony are fallible. But the philosophical skeptic is not concerned, as a scientist would be, with distinguishing the conditions in which these sources are likely to fail from those in which they can normally be trusted.
>
> —A. J. AYER, *The Problem of Knowledge*

A. J. Ayer points in the above passage to a distinction between scientific and philosophical standards for knowledge. His distinction favors the general scientific assumption that highly probable, although not certain, statements are knowledge. That assumption, and the acceptance of science by philosophers as a form of knowledge, did not always hold (if it does now).[1] If the seventeenth-century empiricists had not turned away from the criteria for knowledge required within the context of philosophical skepticism, natural philosophy may never have developed into the sciences of physics, chemistry, astronomy, biology, and the like. Descartes did not accept highly probable statements as knowledge, and that may be partly why his contributions to the physical sciences were often dead ends, superseded by the Newtonian system.

Descartes wanted to ground science on first principles that were known with certainty. I am here concerned with four answers to the question of why he wanted to do this: (1) It was his nature obsessively to doubt everything, although he was compulsively attracted to certainty. (2) He was following a timeless philosophical tradition that still lives. (3) He thought that the challenge of pyrrhonic skepticism demanded it. (4) After Galileo's misfortunes, Descartes' dedication to science motivated him to find a way to link science with religion so as not to incur the displeasure of the Church.

The first answer is too individually psychological to connect the content of Descartes' philosophy with the intellectual life of his time. In chapter 1 I indicated some of the ahistorical conceptual problems with psychological analysis, in the context of feminist criticism. Even if the determinism implied by the psychological interpretation did not deprive Descartes of intellectual agency as a philosopher, it would sever his connections with other thinkers in the historical period—short of a hypothesis of pre-established neurotic harmony (which may be required for any "psycho-cultural" account).

Psychological analysis is not the only path to this kind of ahistoricism. Traditional philosophical interpretations of Descartes are also ahistorical. For example, Jonathan Bennett has made an interesting case that, for Descartes, necessary truths are necessary only because of how God made us. On that reading, God is not precluded from creating truth outside of our understanding, which truth, to us, would not be necessary, true, or even possible. Bennett's view is consistent with both Descartes' derivation of necessary truths about the external world from God in the *Meditations,* and Descartes' remarks about necessity in correspondence with Antoine Arnaud.[2] However, it follows from Bennett's interpretation that if necessity is only psychological or, in philosophical terms, 'subjective,' then God, who within our idea of Him necessarily exists, could, outside of our understanding, will Himself out of existence. It is difficult to believe that Descartes could not have foreseen this possibility, and it is more difficult to see how he could have accepted such an implication in the context of the skeptical framework of his time. For God to be able to will Himself out of existence is even worse than the possibility that He could be an evil demon—the former would be an ontological catastrophe, whereas the latter could be confined to epistemology. Bennett's interpretation makes sense of Descartes' concept of modality in a way that is philosophically convincing now because contemporary philosophers, not being particularly concerned with God, would not be unduly concerned about a God who could will himself out of existence. But given the skeptical framework that Descartes was working against, any interpretation of this type would be ahistorical because it does not provide an understanding of Descartes' thought that would have applied to his time.[3]

Returning to the question of why Descartes wanted certainty, I am going to put the psychological and traditional philosophical answers aside, because they are ahistorical, and spend most of this chapter in discussion of how Descartes was responding to pyrrhonic skepticism. I will then take up the fourth answer, that Descartes' was appeasing the Church. But to begin, we need to reconstruct the doubts that are at issue.

Descartes' Doubts

Descartes introduces three levels of doubt in the First Meditation: doubt about the accuracy of information coming through the senses; doubt about the perspicuity of conscious states of wakefulness as opposed to dreaming, and of sanity as opposed to madness; doubt that what seems to be real exists and that what seems to be necessarily true is necessarily true. The last is doubt about reality and necessary truth, the worst and most philosophically intense level of doubt. Levels one and two can be dealt with by common sense and a restriction of knowledge claims to deductive truth, respectively. But the third level of doubt is the "withdrawal of foundations" of belief that Descartes initially promises his reader. Here, he raises the possibility that the physical world, including his own body, does not exist at all and that the simplest deductive judgments, such as the addition of 2 and 3, may be in error.[4]

In order to balance the strong influence of habitual consent to "long-established customary opinions," Descartes decides to suppose not that a supremely good and true God has created him, along with his (seemingly) perspicuous ideas, "but that some malignant genius exceedingly powerful and cunning has devoted all his power in the deceiving of me."[5]

In the Second Meditation, taking care not to assent to anything that is subject to doubt, Descartes concludes that he can be sure of his own existence as a thinking thing. As a thinking thing, he has a limited but certain knowledge of physical things ("bodies"), not through his senses or imagination but through an understanding of their primary, that is, mathematical, qualities.[6]

Norman Kemp Smith remarks that Descartes refers to a wider intellectual context of skeptical doubt regarding the Second Meditation in his reply to Objection II, which was published along with the *Meditations*. Descartes there wrote: "I have, in years long past, seen several books written by Skeptics and Academics treating of these questions (and though it is not without distaste that I have again served up this stale dish), . . . I had no option save to reserve for them this entire second Meditation."[7] Descartes' weary tone can be taken at face value to suggest that he found skepticism a dull, dead issue; or it could suggest that he found skepticism alarming and meant to play it down as a way of disparaging it. However, Richard Popkin's grounding of early modern philosophy in the issues of skepticism that shaped seventeenth-century intellectual life allows for a more contextualized understanding of Descartes' position on his doubt.[8] I am therefore going to retrace some of Popkin's ground and then use it to get closer to the connection between Descartes' doubt and his interest in protecting science from Church censure.

Popkin's Paradigm and Pyrrhonic Skepticism

Popkin anchors *academic skepticism* in the Platonic Academy of the third century B.C., when it was held that the senses are unreliable, that we have no certainty our reasoning is reliable, and that there is no criterion of truth or falsity for judgments. In this academic tradition, none of our knowledge goes beyond empirical reports of appearances and even our best knowledge is merely probable. Pyrrhonic skepticism began with Pyrrho in Alexandria in the fourth century B.C., but it was first theoretically formulated by Aenesidemus in the first century B.C. In 200 A.D., the pyrrhonic skeptic Sextus Empiricus developed procedures or *tropes* for suspending claims made by both academic skeptics and dogmatists. Whereas academic skeptics had been prepared to assert that knowledge is impossible, pyrrhonic skeptics aimed to live in "quietude," in a suspension of belief about whether knowledge was possible; and they were content to base action on natural inclination, appearances, and social rules, without making judgments.

In the sixteenth and seventeenth centuries, a new awareness of Greek skepticism coincided with Reformation and Counter-Reformation theological controversies, when Sextus Empiricus' pyrrhonic skepticism was rediscovered by Michel de Montaigne, Marin Mersenne, Pierre Gassendi, Robert Boyle, and others. Skepticism was at that time compatible with religious faith because it seemed to relate to knowledge claims about the secular world only. Dogmatism, the view that there is at least one certain, nonempirical or transempirical claim, was the antithesis of skepticism.[9]

During the Reformation, disputes over the "rule of faith" or the standard of religious knowledge occurred in relation to natural knowledge. Luther's denial of Church authority revived the ancient pyrrhonic *problem of the criterion:* Disputes about criteria for knowledge require an accepted outside criterion in order to judge the disputed criteria, but the outside criterion cannot be found until the disputes about the criteria are settled. This circularity was at the heart of the new *crise pyrrhonienne*. Erasmus, in *De Libro Arbitro* (1524), defended Catholicism based on faith from a position that nothing could be conclusively known in theological controversies. Luther responded *dogmatically* to Erasmus' skepticism with his subjective criterion of certainty based on scripture and conscience. Sextus Empiricus' *Hypotoses* and other works were published in Latin, Greek, and English by 1591. From then on, skepticism became independently important as a secular philosophical subject—as Popkin analyzes the Reformation theological disputes, philosophical skepticism had been only indirectly at issue there.[10]

Popkin views pyrrhonic skepticism as the driving force behind seventeenth-century philosophy, but in ways that left beliefs in God unscathed. However, by the eighteenth century, through Hume and the French philosophes, skepticism became a standpoint from which beliefs in the existence of God could be philosophically questioned. That is, although Protestants impeached Catholic Church authority throughout the seventeenth century, and Catholics, in defense of the Church against Protestant dogmatism, were often skeptical about Protestant belief claims, atheism was not intellectually considered by any of these thinkers.

Popkin's interpretations have the puzzling result of treating skepticism as a complex of ideas with a life of its own. Regardless of whether skepticism is explicitly recognized by skeptics for what it is, with an awareness of its historical antecedents, on Popkin's account skepticism moves from philosophy (in the ancient period), to theology (after the Renaissance), back to philosophy (in the seventeenth century), and then into philosophy of religion or religion (in the eighteenth century). Perhaps this interdisciplinary movement expresses a common intellectual disposition, or perhaps it is a sign that the practice of skepticism is a useful polemical tool, depending on how sincere and serious the so-called skeptic is. But Popkin does not suggest what drives the movement of skepticism itself, perhaps because his analysis is not anchored to the individual situations of the skeptics.

Returning to Popkin's account of the late sixteenth century, Gentian Hervet's 1569 edition of Sextus' work was introduced as a cure for dogmatism that would lead to the humility necessary for a quiet acceptance of the Church doctrine of Christ. Francisco Sanchez then developed pyrrhonic skepticism as a philosophical critique of Aristotelianism in his *Quod Nihil*

Scitur in 1576. Sanchez argued for nominalism as well as empirical observation, and his conclusion that knowledge was unobtainable was closer to academic than pyrrhonic skepticism. At the same time, Montaigne, who was far more influential than Sanchez, wrote his *Apologie de Raimond Sebond.* Montaigne used Sebond's arguments to present a fideistic view that pure faith is the foundation of religion. To justify this faith he presented his own pyrrhonic skeptical analyses of the frailties of human reason for philosophical conclusions, the relativity of moral beliefs among cultures, the lack of agreement among scientists, and the failures of sense knowledge. For Montaigne, all of these uncertainties culminated in the problem of the criterion and complete doubt, for which the only solution could be faith.

Popkin argues that it is evident in Montaigne's *Apologie* that skepticism had moved from theology to secular epistemology, and that faith was there offered not merely as a solution to theological conflict but as an alternative to insoluble epistemological problems.[11] However, faith itself, even faith in God, could not fill the lack of a rational justification or foundation for scientific knowledge. The Catholic Counter-Reformers who used skepticism to undermine Protestant claims about the interpretation of scripture themselves needed no defense against skepticism because their faith in Church authority was compatible with pyrrhonic ignorance or suspension of judgment. However, those natural philosophers who took scientific knowledge seriously did not have the same security. Jean Bodin, John Chamber, and Pierre Le Loyer all tried to offer epistemological defenses of sense knowledge between 1581 and 1605. By the 1620s, Fathers Mersenne and Gassendi were using pyrrhonic anti-Aristotelian arguments against alchemy and Rosicrucianism.

Also in the early 1600s, skeptical descendants of Montaigne combined pyrrhonism with anti-Aristotelianism against orthodoxy and traditional authority. Popkin identifies these "free thinkers," the *libertins érudits,* as Gabriel Naude, Guy Patin, Leonard Marande, François de La Mothe Le Vayer, Pierre Gassendi, Samiel Sobrière, and Isaac La Peyrère. Naude and Patin were mainly humanists who were not interested in science. La Mothe Le Vayer and Marande used skepticism to undermine the possibility of scientific knowledge. Sobrière was a disciple of La Moth Le Vayer and Gassendi, and he applied extreme pyrrhonic skepticism to judgments that went beyond appearances in the sciences, especially medicine. Gassendi, however, was a professor of mathematics who revived Epicurus' atomic theory. After his early attacks on Aristotelians and pseudoscientists, he sought a *via media,* between pyrrhonism and dogmatism, for empirical knowledge.

In his positive work, after he had shown that there can be no necessary science of nature, Gassendi accepted experience as the source of all knowledge.[12]

Some of the reactions to these *libertins érudits* are noteworthy. Father François Garasse vilified them as atheists in his *Somme Theologique,* which the Church soon repressed because it had an *entente cordiale* with the free thinkers. There were other Catholic dogmatists who attacked them, but the philosophical responses were (of course) more measured. Popkin divides these into Aristotelians who failed to grasp the problem of the criterion (that any criterion for the truth of conflicting beliefs would itself need to be justified), such as Pierre Chanet, Charles Sorel, and perhaps Francis Bacon, and mitigated skeptics and positive thinkers. Popkin does not think that mitigated skepticism, which eventually became the prevailing scientific view, was accepted until David Hume presented it. However, the philosophy of science of the early fellows of the Royal Society (as will be discussed in chapter 8) would seem to qualify them as mitigated skeptics.[13]

Herbert of Cherburg and Jean de Silhon both tried to combat the nouveau pyrrhonisme with positive knowledge claims. Herbert's *De Veritate,* published in 1624, was an attempt to explain how to arrive at truth through universal "Common Notions" that he referred to as the criterion for breaking the skeptical circle. But as Gassendi quickly saw, there is no way to arrive at universal common notions given existing diversity of opinion. Descartes added that Herbert had failed to say what truth was. Silhon, in ways similar to Descartes, held that in order to know that God exists and the soul is immortal, it is necessary to show that knowledge is possible, that is, to defeat skepticism. But instead of attempting to justify knowledge, he claimed that knowledge must be possible because it exists in logic and the sciences. Silhon presented the *cogito* (that each person can know that he exists), but did not make the effective use of it that Descartes did. (Descartes was friends with Silhon and Mersenne and corresponded with them about skeptical problems; Popkin suggests that Silhon plagiarized unpublished work of Descartes).[14]

Although I have barely scratched the surface in the foregoing outline of parts of Popkin's paradigm, I think it suffices to situate the *Meditations* in the intellectual history antecedent to its appearance. Beginning from the necessary truth of his own existence, which could be any thinker's existence, Descartes offered a proof for the certainty of scientific knowledge, provided it was expressed in mathematics that was already known to be certain. That proof was based on the existence of God, for which Descartes also offered a proof, even though such proofs were not at the time considered to be primary philosophical problems. Descartes addressed the problem of the criterion, as a philosophical problem, by bringing doubt to the third level of the evil demon hypothesis. Popkin's historical account thus goes a long way toward explaining what Descartes was doing in the *Meditations*. However, it does not tell us why Descartes did it—it does not bring Descartes' intellectual agency to light, or to life. Insight into Descartes' specific reasons for writing the *Meditations* when he did would begin to do that.

Descartes' Reasons for the *Meditations*

Popkin's model has worked so far because it is a research program (in Imre Lakatos's meaning for contemporary thinkers).[15] If philosophers who raised skeptical problems can be understood as responding to other philosophers who raised skeptical problems, then their work can be interpreted as part of a long-term skeptical project in the history of thought. However, this does not mean that any particular philosopher who is subsequently interpreted as part of the project saw himself as participating in the project. The history of skepticism is in that sense different from what Charles Schmitt calls the *historiography* of skepticism, which would be the chronology of problems and solutions that were self-consciously posed and addressed as skeptical problems and solutions.[16]

In other words, the intellectual-history explanation of Descartes' doubt in the *Meditations,* in interpreting the work as a solution to skepticism, explains the expressions of doubt within the text. But it is not intellectual biography because it does not explain Descartes' individual relation to those doubts in the text. We don't know that Descartes saw his primary philosophical purpose to be the overcoming of doubt as the skeptics had presented it, or that he took the doubts he wrote about seriously. There is no evidence prior to the *Meditations* that he had doubts about the truth of science as a source of knowledge. Neither did he seem ever to have had religious doubts, apart from the evil demon hypothesis that was used to pose an epistemological rather than a religious problem. Descartes was already deeply committed to science, as well as mathematics, when he wrote the *Meditations*. If his doubt was not wholly sincere but a persuasive (rhetorical?) device to overcome the pyrrhonic doubt adrift at the time, the question is why he wrote the *Meditations* if he already had confidence in science and he already believed in God. I think that there are two lines of biographical response to this question, one psychological and the second professional.

Ben-Ami Sharfstein provides a non-anachronistic, psychological discussion of Descartes' early moral misgivings about the curiosity that led to his anatomical research, especially his dissections of calf fetuses. Scharfstein relates how Descartes thought that curiosity was a moral vice, which if left to itself could be cured only by gaining knowledge. But Sharfstein points out that Descartes also thought that a more detached and comprehensive remedy for this natural evil would be general principles by which everything particular that would otherwise be an object of curiosity could be explained.[17] The formulation and consideration of general principles would slake curiosity without indulgence in the vice.

The *Meditations* would have put an end to Descartes' uneasiness about curiosity because Descartes there claims that first principles, expressed in mathematical form, could be developed for all fields of scientific enquiry. Indeed, at the beginning of the First Meditation, he writes that it would be an "endless labor" to doubt separately each particular belief that could be doubted. Descartes' first principles are known by the understanding, or the mind, which is identical to the (indivisible) soul. God, who is good, is the creator of the human soul, as well as the creator of the external world that the soul's knowledge represents. Therefore, there is nothing morally wrong about this kind of knowledge or its acquisition and contemplation. On the contrary, such scientific knowledge would be the fulfillment for human beings of the divine order. Since Descartes *is* his soul, the *Meditations* in that way absolves him of the sin of his curiosity.[18]

However, this answer does not account fully for Descartes' motivation in writing the *Meditations* because it does not explain why he found it necessary to address pyrrhonic skepticism in that work. He might have simply gone straight to the proof for the goodness of knowledge without raising the ante in skeptical terms by introducing the evil demon. But at the very outset he says, "the withdrawal of foundations involves the downfall of whatever rests on these foundations and what I shall therefore begin by examining are the principles on which my former beliefs rested."[19]

The second line of response to the question of Descartes' motivation for writing the *Meditations* refers to his professional position. In a way similar to Bacon's use of rhetoric to promote science, Descartes used the philosophical problems raised by pyrrhonic skepticism as a foil against which to appease the Church about science. He meant with his rationalist derivation of human knowledge to provide a religious justification of science so that the ecclesiastical authorities would not disapprove of his scientific work, and so they might also use his philosophy as an official pedagogical source. Viewed this way, the *Meditations* intensifies skeptical doubt, which only incidentally becomes doubt about the existence of God. He constructs a fundamentally rational answer to this doubt, and the result is that the conclusions of science can be seen as not only certain, but good, too. Their goodness rests on God's goodness because the certainty of mathematical scientific knowledge is guaranteed by God's goodness, which results from His metaphysical perfection as an infinite substance. If God were less than completely good, He would be less than infinite because vice is something less than virtue, and it is impossible for God to be less than infinite. Along the way, Descartes has provided, through clear and distinct ideas, a rational proof of the existence and goodness of God. But most important is his theological justification for science through the certainty of its first principles: Their certainty ought not to be (cannot be) doubted because they seem to be certain, and God, being good, would not deceive us about such seeming.

All of the biographical summaries that have accompanied the varied editions and translations of his works contain ample evidence that after Galileo's censure Descartes was apprehensive about the Church's disapproval, and that he actively sought ecclesiastical favor. Descartes was born in 1596 and received a Jesuit education, from which he valued his mathematical studies above all. He took a Baccalauréat and a License in law in 1616. In 1618 he traveled with an army at his own expense. In 1619, after Isaac Beekman introduced him to scientific questions, he had a vision of science expressed in mathematics and the three famous dreams (discussed in chapter 1), which he interpreted as a divinely sanctioned revelation of his destiny to unify all of the quantifiable sciences under mathematics. Descartes' early interest in mathematics and his revelations about the mathematical nature of science clear him of having invented scientific certainty in order to pacify the Church.

Descartes traveled throughout the 1620s, came in contact with skepticism and at least once, in the presence of the Papal Nuncio, publicly spoke in favor of certainty for the foundations of the sciences. In 1628–29, he wrote part of the *Rules for Our Native Intelligence,* which introduced the idea of a universal science of quantity. After sun-halos were observed in 1629, Descartes began work on his *Treatise of the Universe* (Le Monde), about meteorology and physics. Before its scheduled publication, he heard about the Church condemnation of Galileo and he suppressed the *Treatise.* He then wrote the *Discourse on the Method* as a preface to the *Essais* and , by 1640, the

Meditations. His concerns about Church censure were well known to his correspondents, and he first published the *Meditations* with six sets of Objections and Replies, with the aim of having the *Meditations* accepted as official Catholic teaching. If the Church had accepted the *Meditations,* as he hoped, his scientific work would have been secure in a way that Galileo's obviously had not been.

There may be more to an interpretation of the *Meditations* as a theologically friendly foundation for empirical science than the matter of avoiding Church censure. There is the question of proof regarding scientific principles that the Church rejected, in particular the heliocentric theory. As Stephen Gaukroger indicates, the heliocentric theory, for which Galileo had been punished in 1633, was already generally associated with mechanistic explanations. Descartes was definitely a mechanist, scientifically, and probably a heliocentrist as well. Gaukroger reads the *Meditations* as providing a transcendent metaphysical foundation for mechanism and, derivatively, for heliocentrism. On Gaukroger's interpretation, in justifying his first principles of science, Descartes would also have justified mechanism with an argument that would have avoided the Church's objections to Galileo's views, which had been rejected because they conflicted with Church doctrine in the absence of compelling proof for their truth.

If Gaukroger is correct in ascribing a project of positive proof for mechanism to Descartes in the *Meditations,* then accepting Descartes' reasoning would have committed Church authorities to reinterpreting those parts of scripture and doctrine that contradicted mechanism.[20] However, Descartes does not explicitly argue for the absolute truth of mechanism or the heliocentric theory in the *Meditations,* and he may have been content with the indefinite connections between mechanism and those first principles of science that he claimed were guaranteed by God. Thus, the *Meditations* may not have represented the offensive strategy implied in Gaukroger's interpretation, but may have been no more than a defensive attempt to ground science on theology in the most general terms.

That several of Descartes' correspondents thought he was overly apprehensive about the Church supports the defensive interpretation. However, many clerics did not accept his answers to skepticism and instead believed that his hyperdoubt had made matters worse for the defense of Catholicism against Protestantism. The Church put Descartes' philosophical writing on the banned list in 1663, which supports Gaukroger's interpretation.[21]

While Descartes met skepticism about science head-on, his colleagues Mersenne and Gassendi were less dogmatic about science and ultimately more successful in advancing its credibility. The same can be said for English empiricists, although they were not under the same religious pressures to prove the goodness of science. After Bacon's false theoretical but hugely successful practical foundation, English empiricists took up the Gassendi-Mersenne solution to philosophical skepticism. Popkin's paradigm will again prove useful in chapter 3, in describing the philosophical connection between the continental and English empiricist epistemology of science. However, as with the discussion of Descartes in this chapter, information about individual career situations will be needed to contextualize further the skeptical paradigm.

Three

The *Via Media* and English Empiricism

> Scientific thought takes its ultimate point of departure from problems suggested by observing things and events encountered in common experience; it aims to understand these observable things by discovering some systematic order in them; and its final test for the laws that serve as instruments of explanation and prediction is their concordance with such observations.
>
> —ERNEST NAGLE, *The Structure of Science*

Like Descartes, Bacon believed that there could be certainly known first principles in science that described truths not accessible to the senses. Unlike Descartes, Bacon held these principles as almost Platonic ideals and directed his epistemological efforts toward encouraging empirical scientific observation and experiment that would lead to the first principles through reflection on inductive generalizations. It was Bacon's utopian enthusiasm about experimental inductivism and not his dogmatic confidence in the certainty of first principles that underpinned English Natural Philosophy, that is, science, as well as philosophy of science, before Newton.

Contemporary historians of ideas have indicated similarities between the continental *via media* and English empiricism but they do not draw the connection as sharply as I would like to in this chapter, perhaps because they tend not to zero in on the common interest in scientific practice among those who addressed the epistemology of science. The continental-English connection determined the early international character of what came to be modern science in the sense of Ernest Nagel's description above.

This chapter has five parts: discussions of Pierre Gassendi's *via media,*

Francis Bacon's philosophy of science; Thomas Hobbes' metaphysics of science, John Locke's inductivism; and Robert Boyle's atomism. Bacon (1561–1628) was born earliest; Hobbes (1588–1679), the most long-lived, knew Bacon, Gassendi (1592–1655), and Locke (1632–1704); and Hobbes' work was eclipsed by Boyle's (1627–1691). However, I do not begin with Bacon because I am more interested in the movement of ideas from the continent to England, where the modern tradition of experimental science became strongly entrenched through the Royal Society. The Royal Society was chartered in the 1660s, and Bacon's influence was not effectively realized before then, though Gassendi's was.

I claimed in chapter 2 that Descartes addressed pyrrhonic skepticism in the *Meditations* in the hope of winning Church approval of science, probably so that he could safely continue his own studies. Descartes was not alone in the strength of his commitment to the practice of science. All of the thinkers discussed here found the new science interesting and valued their involvements with it. Science for them was a recreation taken seriously, a fascinating and enjoyable subject of study and an activity not only worthy of defense, but likely to be pursued even if it could not be justified. For the first time, all at once, gadgetry, experimentation, demonstration (in the sense of spectacle), and shared recorded observations supplemented two-dimensional texts in the real life of adult learning. R. S. Woolhouse discusses the concern of Meric Casaubon, a classical scholar, about the frivolity of the new science. Casaubon ridiculed a story Gassendi told about a friend who observed a louse and a flea fighting under a microscope. Gassendi's friend was said to have learned something that would help him "rule his passions."[1] We tend to read history as though scholars now dead could never have been deliberately silly. But it might not be far-fetched to view the new science as a form of entertainment in its day, as fun.

Gassendi's *Via Media*

Pierre Gassendi and Marin Mersenne were close friends. Both were Catholic priests involved in empirical science and its philosophical justification. The importance of their work has only recently begun to be recognized because the philosophical tradition placed them in the shadow of Descartes and considered their ideas only through their influence.[2] This was particularly true of Gassendi. Mersenne, an energetic propagandist of the new science, encouraged practicing scientists and kept them informed through correspondence, and he secured his own fame by publishing their work, especially Galileo's.[3]

Although the Church's condemnation of Galileo came through the authority of Cardinal Bellarmine (later, St. Robert), during the years preceding the condemnation there was much discussion within the Church about the reality claims entailed in accepting the Copernican system. The crucial issue for Bellarmine was whether accepting the Copernican system required merely making use of a hypothesis that would "save," that is, do no more than not rule out, the astrological "appearances" or whether acceptance of the heliocentric account involved a commitment that it was objectively true. Bellarmine warned Galileo that the Church would not approve the latter view because it conflicted with the geocentric account in the Bible and had not itself been proved. In a letter written in 1615, Bellarmine made his position clear.

> I say that if there were a true demonstration that the sun was in the center of the universe and the earth in the third sphere, and that the sun did not go around the earth but the earth went around the sun, then it would be necessary to use careful consideration in explaining the Scriptures that seemed contrary and we should rather have to say that we do not understand them than to say that something is false which had been proven. But I do not think there is any such demonstration, since none has been shown to me. To demonstrate that the appearances are saved by assuming the sun at the center and the earth in the heavens is not the same thing as to demonstrate that in fact the sun is in the center and the earth is in the heavens. I believe that the first demonstration may exist, but I have very grave doubts about the second.[4]

Galileo did not follow Bellarmine's advice, but I think that Mersenne and Gassendi did in a more general way than Bellarmine's acceptance of the Copernican theory would have required. They thereby provided extensive theological cover for those who continued to work within the Copernican system and for those who found atomism and materialism persuasive. However, it is important to realize that this reading does not commit one to interpreting hidden heretical or atheistic beliefs in either priest. They may have had such beliefs, but it is more likely that, like Bellarmine and also Descartes, their belief in and allegiance to Church doctrine went beyond mere obedience. If the new science could not be reconciled with the Church, Mersenne, Gassendi, and many of their contemporaries would have been in intellectual and spiritual difficulty, besides practical danger from the ecclesiastical authorities. Mersenne's solution to pyrrhonic skepticism was thoroughly pragamatic. Gassendi built on Mersenne's work, not only to justify science within Catholicism, but also to provide a positive theory of scientific knowledge that could be used to organize and rationalize the results of observation and experiment within science.

Unlike Descartes, Mersenne defended scientific knowledge against pyrrhonic skepticism by accepting the skeptics' charge that such knowledge could not be demonstrated with complete certainty. Nonetheless, like perceptual information that was acquired through the senses, scientific knowledge existed in accepted and useful forms—it was "good enough." There was therefore no need to base science on (unattainable) self-evident truths or valid implications from such truths; it did not matter that the skeptics' standards could not successfully be met by the dogmatist.[5]

Gassendi was a professor of philosophy at Aix and of mathematics at the College Royale in Paris. In addition to Mersenne, he knew Johannes Kepler, Galileo, Hobbes, and Descartes, and he published his observations in astronomy, anatomy, and physics. His first important philosophical work was his *Exercises in the Form of Paradoxes against the Aristotelians,* originally prepared as a commentary to his exposition of Aristotle's theory of *scientia* in lectures.[6] These *Exercises* were presented in satire, and their mockery of the Aristotelian tradition that scientific truths had to be deduced from essential definitions was in the spirit of the *libertins érudits* mentioned in chapter 2.

In Gassendi's 1644 *Fifth Set of Objections* to Descartes' *Meditations,* he proposed that there is no way to guarantee that ideas we take to be clear and distinct really are the representations of reality "outside" of those ideas. He thereby raised what Descartes acknowledged to be the "Objection of Objections" to his system, namely that we cannot know whether our knowledge is about anything beyond our ideas. Descartes seems either to have dismissed this possibility out of hand or realized that he had no convincing refutation of it.[7] However, Gassendi developed a philosophy of science that was compatible with the Objection of Objections because it allowed scientific knowledge to be fallible and conditional. This was his *via media* between the pyrrhonic position that we can know nothing and the Cartesian dogmatic position that we can be certain about first principles. Gassendi's *via media* not only protected the claim that science provided knowledge, as Mersenne had done, but in not claiming objective truth for science, it removed it from Church censure on those topics where scientific findings might contradict Church doctrine. Thus, by making concessions to the skeptics about certainty and objectivity, Gassendi made highly effective use of skepticism as a defense against Church censure. Gassendi and

Mersenne were both counted among the *libertins* who were subject to attack by Counter-Reformation fanatics for their secretiveness and stoical worship of nature. Although there was no Inquisition in Paris, the Parlement of Toulouse had condemned Vanini to death in 1519 and in 1622 Fonanier was burnt at the stake in Paris, both for heretical teaching.[8]

In his posthumously published *Philosophical Treatise,* Gassendi came close to anticipating Karl Popper's twentieth-century argument against inductive proofs. Gassendi, like Popper, insisted that inductive generalizations can be falsified by one counterinstance: "It is enough to make your universal declaration false if one single case contradicts you."[9]

Gassendi's positive thesis about what it was warranted to claim based on imperfect scientific reasoning went a step beyond a mere "saving of appearances," which would have been in accordance with Bellarmine's warning—although he clearly meant to stay within the area of hypothesis and explanation—to the avoidance of knowledge claims about reality. The positive thesis emerges in his treatment of the traditional topic of "hidden things" and its relation to Epicurean atomism. According to the ancient Greeks, some matters of fact could never be known and are completely hidden, for example, whether the number of stars is odd or even. Other phenomena are temporarily hidden, for example, objects in the visual field that are blocked by other objects. But there is a third kind of things that are naturally hidden, such as the pores of our skin. Gassendi argued that there can be "indicative signs" of such things, for example, sweat, even though those things themselves will never appear to us. Furthermore, Gassendi claimed, against the skeptics, that in reasoning back to these hidden things we can build up to knowledge of them that will result in useful predictions and explanations—even though we are not able to assert that they exist.[10] Gassendi drew out an extended case study of such hidden things in his development of Epicurean atomic theory. This work passed into the corpuscularian theory of the English empiricists, which will be discussed in the section on Boyle at the end of this chapter. But it was Gassendi, and not Boyle, who was the first thinker in modern science to develop an idea of what later came to be known as 'theoretical entities,' for example, atoms. Gassendi's hidden things, as theoretical entities, were meant to be firmly anchored in empirical studies rather than metaphysical or theological speculation. The empirical anchoring of theoretical entities progressed in England through Boyle's work, and Gassendi's ontological skepticism about theoretical entities was not directly taken up by the scientists and *virtuosi* of the English Royal Society in the seventeenth century.

Bacon's Empiricism

Bacon believed that after empirical generalizations were drawn from scientific observation and experiment, it would be possible to arrive at ultimate causes or "forms" behind the appearances of natural phenomena. Exactly what Bacon meant by these "forms" is unclear because he underestimated the importance of mathematics in the physical sciences, which could have provided a Cartesian or Newtonian translation, and he did not provide examples of his forms. Bacon has been interpreted both as reverting to an idea of Aristotelian essences as formulae for the arrangements of primary qualities and as seeking divine or alchemical truths beyond empirical appearances. For instance, in *Sylva Sylvarum* he mentions seeking the "schematism" of gold so as to produce it artificially. Bacon's contemporaries, like present-day empiricists, had difficulty with his description of ultimate scientific knowledge in Parts IV-VI of the *Great Instauration,* and it was thus put aside. His immediate successors took up his proposals for the methods and practice of science as set out in the *Novum Organum* (and Part II of the *Great Instauration*). These proposals laid out an inductive logic for discovering ordinary causes and effects in scientific investigation, and they endured through John Stuart Mill's revision of Bacon's Three Tables of Investigation into the methods of *agreement, difference,* and *concomitant variation.*

Bacon's vision of the practice of science was first presented in *The Advancement of Learning* and offered in utopian form in "Solomon's House" in *The New Atlantis*. However, the practice of science as a government-funded, systematic, cooperative project to compile *natural histories* was not implemented under James I, as Bacon hoped, but in the generation following his death, under James's grandson, Charles II, who sponsored the Royal Society.

Bacon did not have much to say about continental controversies concerning scientific knowledge or Gassendi's sophisticated *via media* solution to questions of scientific realism. He avoided the details of skepticism except for sharing a widespread impatience with a priori Aristotelian beliefs that knowledge of the external world could be structured by species to conform to logical systems of definitions of essences. He was not particularly interested in theoretical aspects of the Copernican revolution, in mathematics, or in undertaking scientific research himself, although he did write lengthy reflections about his contemporaries' scientific discoveries and observations.

Bacon emphasized research that would result in technological advances.

He was impressed by Marco Polo's contribution to geography, the printing press, gunpowder, and new techniques for refining metals. His ultimate justification for science, which he described as reading nature like a second Bible, was the betterment of human life: "For man by the fall fell at the same time from his state of innocence and from his dominion over creation. Both of these losses however can even in this life be somehow repaired, the former by religious faith, the later by arts and science."[11]

Bacon's notorious worldliness ran from extravagance—he lived lavishly and died in debt—to close shrewdness in advancing his own political position. He was the impoverished youngest son of Elizabeth I's Keeper of the Great Seal. After initial interest in him, the Queen withdrew support when Essex, his patron, was convicted of treason (even though Bacon prosecuted Essex's case). When James I succeeded to the throne, Bacon advanced through the ranks of squire, solicitor general, and his father's post, coming to rest, through the help of friends, and despite a long-term antagonism with Sir Edward Coke, as Lord Chancellor. Throughout his political rise and after his fall due to charges of accepting a bribe in a court case (which may have been related to the Coke antagonism), he published numerous volumes on science and philosophy.[12]

Bacon cast himself as a philanthropic benefactor of mankind. During his lifetime, the word science still evoked the hermeneutic, elitist, leisure-time activity it had been for the ancient Greeks.[13] Perhaps the following excerpt from Bacon's preface to *De Interpretatione Naturae* was true:

> Believing I was born for the service of mankind, I set myself to consider what service I was myself best fitted to perform. Now if a man should succeed, not in striking out some new invention, but in kindling a light in nature—a light that should eventually disclose and bring into sight all that is most hidden and secret in the universe—that man (I thought) would be benefactor indeed of the human race.[14]

But he also wrote that benefit had to be bestowed appropriately, and equality was not to be assumed:

> Neither give thou Æsop's cock a gem, who would be better pleased and happier if he had a barley-corn. The example of God teacheth the lesson truly; *He sendeth his rain and maketh his sun to shine upon the just and unjust;* But he doth not rain wealth nor shine honour and virtue upon men equally.[15]

The idea of knowledge and technology for the practical good of humanity could be viewed not only as a justification for the practice of science, but as a royal motive, supplied by Bacon, for funding scientific projects at a time when the divine and absolute rights of the sovereign were in question. Indeed, Bacon's political rivalry with Coke was partly based on their different philosophies of law. Coke believed that the common law was above all executive power and that the king and parliament together were entitled to rule, subordinate to that law. Bacon, however, believed in an absolute, although not divine, right of the sovereign to rule.[16]

The stability of Bacon's royalism is demonstrated by his obedience to Elizabeth against Essex at a time before his major philosophical and scientific works had been published, unless one argues that it was always in his immediate political interest to support the crown. It is difficult to insist that Bacon's philanthropic view of science was wholly self-serving because the effects of modern science have been of benefit to humanity. In contrast to Bacon, Descartes and Gassendi did not seem particularly concerned with the humanitarian benefits of science, perhaps because the authorities over them were more concerned with how scientific knowledge fit into the Counter-Reformation.

Hobbes' Metaphysics

In the cross-fertilization of French and English empiricism, Hobbes and Locke represent interesting movements back to the continent. Mersenne praised Hobbes for having offered a new science of man in *De Cive,* which added to the evidence against skepticism.[17] Later, Voltaire was to battle the Jesuits on behalf of Locke's epistemology and possible materialism of the soul in his "Letter on Mr. Locke."

As a young man, Hobbes visited Bacon during the years after his fall. Like Bacon, Hobbes had little use for Aristotelian philosophy and was impressed by the practical knowledge acquired through science, specifically Euclid's geometry, Galileo's studies of motion, and Harvey's discovery of the closed circulation of the blood.[18] Hobbes' moral and political theories may have appeared to be a human or social science to Mersenne, based on their geometrical presentation. But Hobbes' concept of science itself, as a philosophical theory of "matter and motion," was very different from what Bacon meant by the particular sciences as sources of empirical "histories" of the world.

Hobbes was in disrepute with the English Protestant establishment due to the atheistic tones of his materialism; in 1688, several bishops in parliament suggested he be burnt as a heretic.[19] Also, his support of absolute monarchy probably motivated more democratic political theorists, such as Locke, to disassociate themselves from *The Leviathan*.[20] Hobbes' long-term exclusion from the Royal Society seems to have been the result of these politics, as well as personal disputes, but it could more justly have been based on his rationalist epistemology, which, although it reduced everything known to the scientific concepts of matter and motion, failed to make a connection between uses of these concepts and actual, scientific observation and discovery. Nor did Hobbes suggest that such a connection ought to be made. Seth Ward, writing on natural theology after Hobbes' 1651 publication of *Leviathan,* accused Hobbes of injuring mathematics in his attempts to reason with demonstrative certainty.[21] Hobbes believed that political science could be derived from physics, through ethics. In his early *Little Treatise* and, later, in the beginning of *De Corpore,* Hobbes formulated a theory of matter in motion that in principle could account for all areas of human experience. The real properties of objects, he theorized, are extension, shape, and motion, and their secondary qualities are caused by the motions of objects in contact with sense organs; sensations, thus caused by external motion, cause human action or behavior.

Hobbes assumed a congruence between the structure of the world and the structure of knowledge when he wrote as though the objects of abstract knowledge, of geometry, were also part of the world. In what was probably a confusion between how things are known and their ontological status, he thought geometry to be a science of motion because its definitions require an understanding of the motions required to construct geometrical figures. According to Hobbes, geometry makes possible both the motions of bodies and the effects of these motions on one another. He therefore reasoned that physics, or the motions of particles within bodies, could be derived from geometry.[22] Again, Hobbes seems to have assumed, in rationalist and Aristotelian fashion, that what is true of the knowledge of the world is true of the structure of it, and vice versa.

According to Hobbes, causation always involved the motions of contiguous bodies, which excludes action at a distance—Newton's concept of gravity, for example. Knowledge of effects can be deduced from knowledge of causes. However, knowledge from effects to causes, such as Galileo's mechanics, is less certain and requires a method of analysis that divided the effects into parts, followed by a synthesis, or understanding of how the parts fit together. R. S. Woolhouse points out that this was Galileo's *reductio-compositive* method, and it is not surprising, given Hobbes' admiration of Galileo, that he adopted it.[23] In discussing the reductio-compositive method, Hobbes made a distinction between limited science or particular scientific inquiries, and indefinite science, which seems to be the general conceptual framework about matter and motion within or according to which particular inquiries are conducted.[24] Hobbes appears to have been working within indefinite science and reasoning from cause to effect, which is why, and how, his philosophy of science fails to be empirical.

Hobbes is commonly reported not to have had any direct interest in the particular sciences, and he conducted no scientific research himself—except for his attempts to square the circle. He had a wealthy patron from the beginning of his career and devoted his long life to private intellectual pursuits with occasional sharp controversies. He was not politically engaged, as both Bacon and Locke were. John Aubrey, the seventeenth-century biographer who wrote more and better about Hobbes than any of his other subjects, reports that Hobbes was proud of not reading much.[25] It is ironic that given his wide circle of intellectual acquaintance, Hobbes seems not to have grasped the main issue for English empiricism, which was the importance of induction, and that given his lack of political engagement, his major influence has been as a political theorist.

Locke's Empiricism

Locke was as deft as Gassendi in presenting a justification for the new science that (almost) allowed him to steer clear of religious controversy. The telling controversy for Locke, as will be discussed in chapter 7, was the question of the materiality of the soul, the same question that won Hobbes so much abuse.[26] Locke also constructed an epistemology that was deliberately neutral about metaphysical ontological questions, especially those involving substance—he claimed that the real nature of matter is as opaque as the real nature of mind.

Locke's epistemology supported empirical scientific practice generally, and it also provided a way of thinking about the human condition that undermined the doctrine of the divine right of kings, to which Locke was politically opposed in theory and action. His two most important works, *An Essay Concerning Human Understanding* and *Two Treatises of Government,* were published almost simultaneously. The *Two Treatises* appeared anonymously, after decades of study, reflection and revision conducted with the kind of discretion, if not outright secrecy, necessary for him to thrive as a hands-on political theorist during the time from the Restoration in the

1660s to the Glorious Revolution that put William and Mary on the throne in 1689. Toward the end of his life, Locke held weighty public office as Commissioner of Appeals and of the Board of Trade and Plantations.[27] The connections among Locke's epistemology, political science, and political and scientific interests will be outlined in this section, and further aspects of the "Lockean paradigm" will appear in almost every chapter after this.

The first book of Locke's *Essay* is a series of arguments against the position, presumably but perhaps erroneously ascribed to Descartes, that there are innate ideas.[28] The position is complicated because Locke does not clearly distinguish between ideas and beliefs and between being born with an idea (or a belief) and becoming aware of it later, with or without the benefit of experience. However, the rest of the *Essay* develops claims that all ideas are based on experience, and that beliefs are based either on experience or reflection about experience. The *First Treatise* is a series of arguments against Robert Filmer's attempt in *Patriarcha* to justify the divine right of kings via an analogy with the power and authority of fathers in families, and a religious claim that seventeenth-century kings were directly descended from Adam.[29]

Locke argues empirically against both innate ideas and the divine right of kings: there is no evidence for universally held innate ideas and no way to trace the lineage of contemporary rulers (not always male) to Adam. Also, as already mentioned in chapter 1, Locke pointed out that governing is different from parenting because parenting is temporary and citizenship is life-long. Locke's positive theses are that all knowledge is built up from information that comes to us through the senses, and that political power can be justified only if it is based on the consent of citizens.[30] In a general sense, the absence of innate ideas would seem to preclude any innate qualifications of a sovereign to rule. But Locke did not use this argument in the *Second Treatise,* either because it would have jeopardized the anonymity of that work or because it raised egalitarian political principles that he did not want to endorse.

Locke's claim that knowledge comes to us through the senses in simple ideas and is built up through combinations of these ideas and reflections upon them allows the collection of data in science, and the organization of that data and conceptual analysis of it, to count as knowledge; it legitimizes science. According to Locke, all that counts as knowledge is empirical in this way. Although he insisted that human knowledge is limited, he also insisted, like Mersenne, that it was "good enough," suited to our condition and useful for living out our lives as the beings God made us. Locke says at the beginning of the *Essay* that after discussions with friends about a different matter, they became perplexed and fell into doubt. He thought that they had taken "a wrong course and that before going further it was necessary to examine our abilities and see what objects our understandings were, or were not, fitted to deal with."[31] One of the things we cannot know anything about is substance and throughout the *Essay* Locke maintains a neutrality about what 'substance' is, about whether the human soul is material and about whether matter can think.[32]

Locke's principal patron during the years in which he composed the *Essay* and *Treatises* was Lord Ashley, earl of Shaftesbury and, for a while, Lord Chancellor. Locke was Shaftesbury's physician and secretary. He collaborated with Shaftesbury on political theory and served him in still-unknown ways in the attempt to keep the Catholic Duke of York, Charles II's younger brother, from the throne.[33] Like Bacon and Hobbes, Locke had little use for Aristotelian philosophy. Unlike Bacon, he did not need to make a case for the use of science or its funding, perhaps because the Royal Society, of which he was a member from 1668 on, already existed. Even though Locke was not a scientist but a *virtuoso* member of the Royal Society, in addition to his practice of medicine for Shaftesbury and other friends, he was familiar with contemporary scientific methods and discoveries.[34] Unlike Hobbes, he had access to much of Boyle's completed scientific theories and research, which must have made it easier for him to classify atomism as a theory of the particular sciences, rather than a metaphysical truth. Although Locke thought that Boyle's atomic or corpuscularian theory was true, he was careful to avoid knowledge claims about the cause of perception and sensation by insensible particles of matter: "We are so far from knowing what figure, size, or motion of parts produce a yellow colour, a sweet taste, or a sharp sound that we can by no means conceive how any *size, figure, or motion* of any particles can possibly produce in us the *idea* of any *color, taste,* or *sound* whatsoever; there is no conceivable *connexion* betwixt the one and the other."[35]

Atomism from Gassendi to Boyle

Walter Charlton formally brought Gassendi's atomic theory into English empiricism with the 1646 publication, in English, of his *Physiologia Epicuro-Gassendo-Charltonia, Or a Fabrick of Science Natural, Upon the Hypothesis of Atoms, Founded by Epicurus, Repaired by Petrus Gassendus, Augmented by Walter Charleton*. Before I discuss this transmission, the different contemporary positions on the truth status of the atomic theory need to be distinguished.

Gassendi held that all matter is probably made up of atoms, though we cannot know this with certainty because atoms do not appear to us. Hobbes held that the material world is made up of atoms and that our knowledge of this is certain and can provide a standard for explaining natural phenomena in the particular sciences. Boyle held that the subjects of physics, chemistry, biology, and other inquiries into gases and fluids that were on his curriculum were made up of atoms. Boyle also wrote numerous Baconian-type histories of experimental support for the atomic theory across these fields. Locke believed that Boyle's theory was probably true, although we could not yet apply it in sufficient detail to explain perception and might never be able to do so. Newton assumed that Boyle's theory was true, and that given sufficient microscopic power, it would be possible to see the largest atoms or corpuscles.[36]

Gassendi and Locke were working within an epistemological philosophical framework that allowed for scientific knowledge as we now understand it, such that, within the particular sciences, atomism would be part of that knowledge. Hobbes assumed the truth of atomism as a general fact about the world that need not be anchored by findings in the particular sciences. Newton and Boyle thought that atomism was true, from within the standpoint of the sciences that they knew, even though they had different theories of what this truth meant: It was hypothetical for Boyle but ontologically absolute for Newton. The philosophical epistemology of Gassendi and Locke could have withstood disproofs of the atomic theory. Such disproofs would have collapsed Hobbes' philosophy but in principle done no more damage to Boyle and Newton, who were working from within the empirical sciences, than given them cause to rethink conclusions that were in principle based on observation.

Carleton's main concern with Gassendi's atomism was to purge it of remaining traces of Epicurean atheism and materialism concerning the soul.[37] But further scientific modifications of Gassendi's views were necessary before they could be incorporated into the consensus of the members of the Royal Society, and this is what Boyle accomplished. Boyle argued for atomism based on the scientific observation of objects that were accessible to the senses. Maurice Mandelbaum calls this kind of inference from observation to what cannot be observed "transdiction," and he makes a plausible case that transdiction became an accepted form of inductive reasoning through Boyle's efforts on behalf of the new science.[38]

Gassendi's hypothetical atoms, which he described as finite in number and directed by God, had perceptible qualities and shapes. Gassendi postulated their existence as a plausible hypothesis, suggested by empirical observation and useful for both explaining and predicting further observation. But Gassendi refused to ascribe mathematical qualities to the divinely instigated motions of atoms, perhaps because that would have involved him in Cartesian certainties about the general nature of the universe, which he thought the pyrrhonic skeptics had demonstrated to be unknowable.[39]

Boyle shared Gassendi's belief that the existence of atoms was based on empirical observation and that they could be used both to explain and to predict such observation; he was also prepared to argue that atoms exist. Mandelbaum reconstructs several of Boyle's main arguments for inferring the existence of atoms by the use of *transdiction*. Boyle began, with reference to recent telescopic and microscopic findings, by pointing out how limited our senses are. He suggested that analogy could be used to extend sense knowledge. For example, in explaining how compression results in increased firmness, the making of a snowball could be described. The result of this method of thinking is that corpuscles or atoms can be described as having qualities similar to objects that can be sensed. Atoms also have the same principles of action as sensible objects and can be explained in mechanical terms by the transference of motion. Boyle's initial justification for this "extension of analogy" is the resulting clarity of explanation, but he went on to supply case after case of confirmed predictions about gases, solids, and heat that were based on the premise that atoms existed.[40]

Boyle was the fourteenth child of the First Earl of Cork, who was the richest man in England. In addition to his work in natural philosophy, Boyle also studied medicine. He was concerned that interest in science not lead to a neglect of religion. Boyle's endorsement of the corpuscular theory was enabled both by the religious tolerance he enjoyed and the private wealth that he continually used to finance scientific research. Given his own interest in science, there was no reason for him not to take it up as enthusiastically as he did. In *The Christian Virtuoso* (1660) he argued that there is no inherent conflict between science and religion.[41] In *A Disquisition about the Final Causes of Natural Things* (1688), he argued that in day-to-day work, a scientist need only be concerned with the primary qualities of particles, even though he was unwilling to dispense completely with the concept of final causes in science. Boyle's public support of the connection between Christianity and science is usually read as a defense of science, but it remained external to the content of science, unlike Newton's religious philosophy of science (as will be discussed in chapter 9).

Boyle's use of transdiction was important in the context of skepticism and English empiricism because it enabled seventeenth-century scientists to institutionalize uncertainty, and even mystery, within accepted scientific knowledge. Reasoning from the known to the unknown and accepting the existence of the unknown on rational grounds accomplished two gains for science: First, empirical scientists acquired the freedom to go beyond experience; second, they were not bound to absolute knowledge claims beyond experience.

In chapters 2 and 3 I have begun to provide a positive account of the intellectual contexts of Descartes, Gassendi, Bacon, Hobbes, and Locke, along with the beginning of a subtext about the value they placed on science and their own careers. This introduction to their "career politics" is anchored in the general Nietzschean assumption, discussed in the Introduction, that individuals seek to increase their own power through life choices and projects, and it fills in some of the gaps left by the feminist critique discussed in chapter 1. In Part II, further aspects of the historical identity of these scientific thinkers will be developed, namely, their marital status, philosophical theories of identity, religion, and peer relationships. Part II will end with Newton's cosmology, in a profile of scientific identity that instantiated some of the worst traits critics have ascribed to male scientists.

Part Two

The New Identities

Four

Bachelors in Life

He that hath wife and children hath given hostages to fortune, for they are impediments to great enterprises, either of virtue or mischief. Certainly, the best works, and of greatest merit for the public, have proceeded from the unmarried or childless men, which both in affection and means have married and endowed the public.

—FRANCIS BACON, "Of Marriage and the Single Life"

Descartes, Hobbes, Boyle, Locke, and Newton were lifelong bachelors. They all lived past sixty, advanced old age for the seventeenth century, except for Descartes, who died prematurely at fifty-four while in service to the Queen of Sweden. Bacon married a girl of fourteen after a three-year engagement; they had no children and were together for fifteen years, until his fall, when they separated.[1] He wrote the above lines a year after his marriage, and there is nothing in the rest of the essay they come from to suggest that men are personally improved by marriage, except, backhandedly, that single men may be "more cruel and hard-hearted (good to make severe inquisitors), because their tenderness is not so oft called upon." Bacon nowhere suggests that marriage could make a man happy, apart from its usefulness: "Wives are young men's mistresses; companions for middle age; and old men's nurses."[2]

My use of the bachelorhood of seventeenth-century philosophers of science as a trope for the connection between science and a development of male gender identity during that period needs to be explained and justified. In this chapter I will situate their bachelor status historically and in relation to their professional lives. I keep a skeptical distance from the emotional aspect of their bachelorhood because we do not know if the generalizations made about human emotions in later centuries can be applied to seventeenth-century affect, due to the different circumstances of personal relationships then.

Bachelorhood and External Social Conditions

Since the 1960s, historical scholarship on the European family between the late medieval period through the eighteenth century has increasingly relied upon a 'primary record' of letters, diaries, wills, biographies, domestic manuals, and local church and court records. It is unknown how incomplete this record is, and much of it is subjective because it was set down incidentally and therefore has to be interpreted qualitatively as narrative. Nevertheless, several informed descriptions of marriage, family custom, and social class influence have emerged that are relevant to the present discussion of bachelorhood.

Lawrence Stone suggests that in the 1500s the Reformation and economic change combined to undermine a centuries-old family structure that was based on extended relationships among core family members, kin, servants, and neighbors, in households that were more self-sufficient than in later centuries. There was a weakening of church and community influence on the family throughout the seventeenth century, despite periods of intense religious enthusiasm, especially during the English Civil War. By 1700, arranged marriages were giving way to independent choice of marriage partners, resulting in stronger conjugal emotional bonds. Previously harsh and dutiful relations to children softened to kindness and concern for the child's well-being. Members of middle- and upper-class nuclear families began to value physical privacy and refinement in personal hygiene and eating habits.

The 30–50 percent rate of child and young adult mortality until the end of the eighteenth century resulted in scant confidence that any particular infant would survive into childhood or child into adulthood. It was unusual for a spouse to live beyond twenty years of marriage. Stone concludes that spouses and children were in principle interchangeable, given the probabilities of death, and that strong emotional ties within the family did not generally develop under this threat of imminent separation. Stone's conclusion does not entail literal interchangeability—there is no evidence of spouse or child swapping—but rather serial replacement of spouses and children, which suggests that relationships within families were based more on roles than on individual personalities.

The changes in the late seventeenth century toward stronger emotional bonds, which Stone calls "affective individualism," took place primarily within the middle classes, which set trends for the rest of society: the upper classes retained arranged marriages to control the inheritance of property; and the poor, who always had more freedom in marriage choice, were slower to institutionalize emotional bonds within the family.

Marriage was traditionally a functional arrangement, motivated by improving, or at least not lessening, financial worth and social status. Romance was not a serious factor, and it was expected that husbands and wives would get used to one another physically, no matter the degree of sexual attraction. There was a liberalization in sexual mores through all social classes during the seventeenth and eighteenth centuries, although the nobility, having always been less self-denying, was quicker to acknowledge the value of sexual pleasure.[3]

In families with property, the rule of primogeniture often left younger sons too poor to marry and set up independent households. When younger sons from families of means did marry, they often could not afford to do so on the same economic (and therefore social) scale in which they had been reared.[4] However, primogeniture itself could not have caused the bachelorhood of all of the philosophers of science under discussion. Bacon, Descartes, Hobbes, and Boyle were all younger sons from the middle and upper classes. But Boyle's father left him the manor of Stallbridge in Dorset and a yearly income of £3,000 from rents in Ireland, so he could have married and lived grandly had he wished.[5] Descartes inherited an independent income, as well as the right to use "Des" in front of his name, from his father, who was a member of the *petite noblesse,* and he also could have afforded to marry.[6] Locke was an elder son and his father was a lawyer and small property owner. He eventually achieved comfortable means and it is not clear that he remained single for economic reasons.[7] Newton was an only child except for stepsiblings; his father, a yeoman farmer, died before he was born. He seems always to have been poor until late in life, but there is no indication that he ever wanted to marry.[8] Hobbes was in service to a noble family all his adult life and, while he could afford to have an emanuensis when palsy made it difficult for him to write, his studies combined with his obligation to his patron may have precluded marriage.[9]

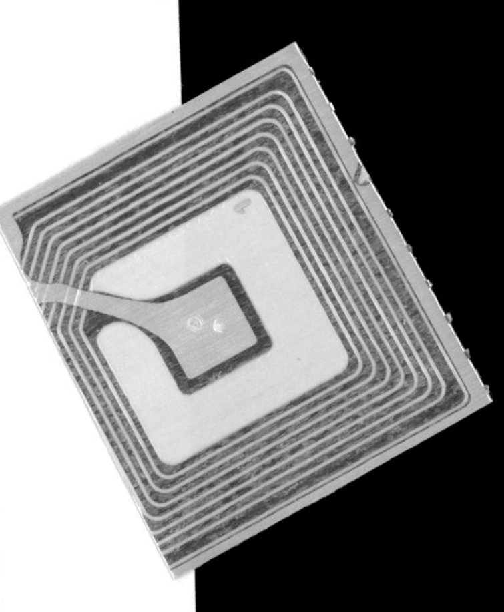

In seventeenth-century England, between 20 and 26 percent of all younger sons never married. Those who did often had to wait until they had accumulated sufficient capital or income to support their families, so they tended to marry in their thirties, about ten years older than heirs. Marriage was also a way of securing patronage from fathers and fathers-in-law, who as heads of households wielded considerable economic power. Not only did fathers have power over their own children, but they also received the dowries of their daughters-in-law, which they often used for their own daughters' dowries.[10]

Although it was more difficult for younger sons than heirs to marry, the majority of men did marry in the seventeenth century. Among the poor, only 10 percent remained single.[11] Aside from Bacon, who did marry, the English philosophers of science seem not to have been overly concerned with social rank or personal wealth. Financial restrictions and a relatively high rate of late marriage might have made the single state of any man of modest means from a privileged class understandable to his peers. However, this does not explain why most of the seventeenth-century philosophers of science never married. In the absence of inheritable property, we do not know if marriage was desirable to most middle- and upper-class men in the seventeenth century. We do know that marriage did not represent personal fulfillment or happiness, if anything did in the way we understand those concepts today. At best, marriage could be a practical benefit for a man. Seventeenth-century men did not generally believe that association with women enhanced their manliness. In close association, a woman could influence a man's actions and make him inferior to other men if he thereby became similar to her. In contrast, association with other men in manly activities increased masculinity.[12]

More specifically, sexual relations with women were believed to be debilitating to men. According to commonly accepted medical opinion, brain tissue, bone marrow, and semen were all of the same stuff. Because ejaculations were believed to draw brain tissue down the spine and out the penis, the preservation of life and reason required that sexual orgasms be limited. It was believed that women were sexually insatiable and men (presumably) had little power to resist female seduction.[13] At a time when the average life expectancy was thirty-two, a relatively celibate bachelor who lived past sixty, especially if he was engaged in intellectual work and related activities, might have credited his longevity and active intellect to his single state.

There were, of course, exceptions. According to Aubrey, Hobbes wrote love verses such as the following after the age of ninety:

Yet I can love and have a mistress too
As fair as can be and as wise as fair,

And yet not proud, not anything will do
To make me of her favor to dispair.[14]

Furthermore, there was a countervailing medical tradition, going back to Galen, which held that complete sexual abstinence was injurious to health. Stone quotes the physician Tomas Cogan, who wrote in 1589:

> the commodities which come by moderate evacuation thereof [semen] are great. For it procureth appetite to meat and helpeth concoction; it maketh the body more light and nimble; it openeth the pores and conduits and purgeth phlegm; it quickeneth the mind, stirreth up the wit, reneweth the senses, driveth away sadness, madness, anger, melancholy, fury.[15]

The demographics of family mortality and rate of bachelorhood were about the same in France during the seventeenth century, except for the stronger influence of the Catholic Church. Priests, of course, could not marry, so Mersenne and Gassendi remain outside of this discussion. Primogeniture also disenfranchised younger sons in France, but it has already been noted that Descartes was financially independent. Since philosophical studies were traditionally under the control of the Catholic Church, it is tempting to suggest that the bachelorhood of philosophers of the new science was the result of this strong tradition of priestly scholarship. However, this would not apply to the Protestant philosophers, none of whom were clerics, in a country where clerics themselves were not only free to marry but encouraged everyone else to do so for moral and religious virtue.[16] Most of the eighteenth-century French philosophers were either free-thinkers or atheists, who, as Robert Darnton points out, tended not to marry either, usually because their living conditions were very unstable and they could not afford to support families.[17]

The Catholic Church had made marriage a sacrament as recently as 1439, and church marriage was made a requirement for legal marriage in 1563, as a tactic against the Reformation.[18] In England in 1653, during Cromwell's rule, a Puritan Parliament restricted lawful marriage to solemnization before a Justice of the Peace.[19] Throughout the seventeenth century, both civil and church weddings were legally binding, as were quick marriages by those under twenty-one who did not have parental permission, and spoken engagements followed by consummation. Eventually, Lord Hardwick's Marriage Act of 1753 required church wedding for legal marriage, although this law was mainly for the purpose of protecting the property of heirs by the use of official contractual ceremonies with witnesses.[20] Overall, religion seems not to have been a primary motive for remaining single during the early modern period. The connection of marriage with churches was an important political tool used to protect ideology and family property, and no one outside of the Catholic clergy was encouraged to remain single. Of course, there may have been individual reasons for not marrying that were connected with personal religious beliefs, but the important point is that there seems to have been no widespread religious support for bachelorhood—or for "spinsterhood."

Bachelorhood and Profession

The bachelorhood of the seventeenth-century philosophers of science seems to result more from their particular career histories and general professional commitments than from social factors external to their work. This is a good place to clarify terminology and explain why Boyle and Newton, who were scientists, are being discussed as though they were philosophers. Today, after more than three hundred years of chemistry and physics proceeding from the theories in those sciences developed by Boyle and Newton, we refer to them as scientists. But in the seventeenth century, Boyle and Newton were natural philosophers. The term *science* still referred to a body of certain knowledge, such as geometry, when it did not mean abstract and empirically irrelevant Aristotelian *scientia*. This is the literal meaning of the platitude that the sciences originated in philosophy and split off from it. In the seventeenth century that split had not yet been accomplished, so insofar as they made philosophical arguments and assumptions about their empirical studies, Boyle and Newton were philosophers, generally, philosophers of science, specifically, as well as scientists who were at the time called natural philosophers.[21]

The bachelorhood of the seventeenth-century philosophers intersected with their work to give them a status that was seriously recognized and respected by the most powerful figures in their society. Noble patronage seems to have been attainable by those who were already philosophers or in a position to become philosophers: Bacon was in service to the earl of Essex (1588–1601); Hobbes had the patronage of William Cavendish, second earl of Devonshire, and his family (from 1608 onward); Locke was adviser, family physician, and secretary to the earl of Shaftesbury (1667–75) and confidant of William, Prince of Orange, in Rotterdam, before William and Mary were crowned. All, at the high points of their careers, were recognized by reigning monarchs: James I made Bacon Lord Chancellor,

Baron Verulam, and Viscount St. Albans (1616–20); Descartes was summoned from Holland to Paris in 1648 to receive a royal pension (which he never collected); Hobbes had hopes of getting royal funding to endow a free school in Wilshire, with a salary for himself as schoolmaster (1665); Boyle was well received at Court at the Restoration and was appointed to the Council of the Royal Society (1663) (although he declined both its presidency and the provostship of Eton because he did not want to take oaths [1662–63]); Locke was appointed commissioner of the Board of Trade (1696); Newton was appointed warden of the Mint (1697) (though this was belated because he was fifty-five and already internationally recognized, and it required the influence of his friends, Locke and Charles Montague, earl of Halifax). These posts and honoraria have always been mentioned in the official accounts of these men's lives. However, except for Bacon's political involvements, they are usually presented as effects or evidence of intellectual excellence, with no special significance as events in the careers of men who wanted to get on in the world and were aware of strategies for doing so.[22]

In English society, which was otherwise ruled by patriarchs through their familial patronage, apart from Bacon's and Locke's political efforts, it is unlikely that such success would have occurred without the development of the new science. Politics offered another way to succeed that was independent of marital status at that time. Bacon had put together the components of his political rise by the time of James I's accession, when he became engaged, and his father-in-law was a commoner (a squire) and not able to help him significantly; Locke was politically active and successful throughout his career, and he was always single.

Love and sex were not primary motives for middle- and upper-class marriage in the seventeenth century, but money and status were. These men probably did not marry because they knew they could do better in the world by remaining single and concentrating their financial resources and social efforts on activities associated with their work. Another reason to marry, of course, was procreation. However, without great property to bequeath or widely accepted beliefs in the intrinsic value of children, there is no reason to think parentage would have been an important goal. Bacon has a distinctly nonbiological opinion on procreation:

> The perpetuity by generation is common to beasts; but memory, merit, and noble works are proper to men; and surely a man shall see the noblest works and foundations have proceeded from childless men, which have sought to express the images of their minds, where those of their bodies have failed: so

> the care of posterity is most in them that have no posterity. They that are the first raisers of their houses are most indulgent towards their children; beholding them as the continuance not only of their kind but of their work; and so both children and creatures.[23]

Thus, Bacon thought that intellectual work was a more cultivated version of biological reproduction and that if the work were original or foundational, attachment to it would be stronger than if one were not "first raiser" of that house (by "house" he would have meant a line of descendants).[24] Bacon's idea could very specifically be applied to Descartes, who deeply mourned the death of his illegitimate daughter but produced his most important work after her death.[25]

Boyle might seem to provide an exception to the hypothesis that, given ambition, the new science was sufficient for success in the world without marriage. Boyle was always rich, he had a long-term interest in science and experiments, and he had enough property to make his will a subject of considerable interest to his biographers.[26] But he never married. Why was he so interested in science if he did not need it to succeed in worldly terms, and why did he not marry to have an heir? If science were a way to succeed, this does not mean that the seventeenth-century philosophers of science were primarily interested in it for that reason; it has already been shown in chapter 2 how much of their work in other philosophical areas may have been undertaken to protect their interest in science. The thesis that science was a way to succeed in the world is compatible with a claim that those who did so succeed had a genuine interest in the subject to begin with. Also, Boyle took care, confirming Bacon's thesis, to leave parts of his estate for educational endowments (the executors eventually targeted funds for Harvard University) and missionary projects to further his religious philanthropy.[27]

Nevertheless, of all the men under consideration, Boyle's life would sustain an interpretation of his single state based on later modern reasons. His health was poor, and after a stroke at the age of forty-two he lived twenty-two years with his favorite sister, Lady Ranelagh, who predeceased him by only one week.[28] These are later modern reasons because they could support a psychoanalytic biographical interpretation of Boyle's bachelorhood that his illness gave him an "excuse" not to marry and to become dependent on care from his sister, of whom he was "overly fond."[29] But why would Boyle need an excuse not to marry in a culture where, rich as he was, he did not have the responsibility of passing on the bulk of the family fortune and had nothing to gain in the world by marriage? And how could it be determined that he was "overly fond" of his sister when there were no standards of fondness for wives to compare his alleged feelings for her? If this sounds obtuse, it may be that we assume too quickly that people in the seventeenth century had the kind of selves (personal identity) that would support the emotional psychology that at its best has been considered normal and natural in the twentieth century. Educated people have assumed since Freud that the human self is fundamentally made up of needs and desires for close emotional attachments, both unconsciously and consciously, if the person is psychologically "well." (Of course, such an assumption cannot be casually waived, and in the next chapter I will begin to question its relevance to the seventeenth century, more systematically, through Locke's philosophical theories of personal identity.)

From his early twenties on, Boyle pursued scholarly, scientific interests. By the time he went to live with his sister, his scientific reputation was securely established. Her house in Pall Mall was close to his scientific colleagues in London. He kept a laboratory there for his numerous experimental projects, and his sister was interested in his work and very well connected socially and politically, so that she could look after his interests generally.[30] Thus, there were ample professional reasons for Boyle to live with his sister after his health failed, and a psychoanalytic interpretation would at best make his choice "overdetermined."

Locke also retreated to the home of a supportive female companion, Damaris Cudworth, Lady Masham, during his final years. Biographers believe he had considered marrying her decades earlier, and his vitriolic opponent, John Edwards, referred to him as "the governor of the seraglio at Oates" during this period. But Locke moved into the household on good terms with Sir Francis Masham, Damaris's husband, and he insisted on paying £1 rent weekly. He brought with him his library of 5,000 books and various other personal effects, which he inventoried on a list that Lady Masham signed (a normal practice for Locke in his frequent changes of domicile). The country air in Essex was better for Locke's respiratory problems than London, and he worked steadily at Oates, receiving visitors and keeping up an active correspondence until his death. The arrangement was clearly conducive to the orderly last thirteen years of his life.[31]

This discussion would not be complete without a broader look at the relationship between bachelorhood and the profession of philosophy. During the early modern period, bachelorhood was required of all university or college fellows at Oxford and Cambridge. Out of the group considered here, Locke and Newton held such positions until middle age, but there is no evidence in either case that this requirement was the cause of their lifelong bachelorhood: Locke could have married when he lost his university post and Newton after he joined the Mint. The required bachelorhood of priests on the continent and academic fellows in England could strengthen an argument for the general association of bachelorhood with philosophy in the early modern period. It could be argued that the prescription of bachelorhood was internalized by philosophers with results that made them different psychologically from men in other occupations. However, I have been making the wider cultural case that early modern marriage did not have the personal value it acquired later.

Ben-Ami Scharfstein has researched the lives of successful philosophers in search of a common psychological and demographic profile. Sharfstein presents a table of ages of parental loss, fields of interest, mental illness, and marriage age for Montaigne, Hobbes, Descartes, Pascal, Spinoza, Locke, Leibniz, Berkeley, Voltaire, Rousseau, Kant, Hegel, Schopenhauer, J. S. Mill, Kierkegaard, James, Nietzsche, Santayana, Russell, Wittgenstein, and Sartre. Some of Sharfstein's findings seem relevant to what has already been covered in this chapter: in only six instances were both parents alive when the philosopher was fifteen; only eight of the philosophers married, and Russell (whose divorces rate support Sharfstein's thesis) and Sartre were the only ones who married during their twenties. However, parental loss and bachelorhood decreased in the eighteenth century, beginning with Berkeley, whose parents survived into his fifties and who married, although not until he was forty-three.[32] This suggests, against Scharfstein's eagerness to connect early parental loss with both high intellectual accomplishment and attachment difficulties, that historical conditions first need be taken into account. Stone notes, for example, that in the mid-seventeenth century a third of the aristocracy had lost one parent by the age of fourteen, and among the poor the rate of orphanage was over one-half.[33] The connection between parental loss and bachelorhood via a psychological deficiency in forming close relationships is on even shakier ground in the twentieth century. Bertrand Russell, for example, lost both parents by the age of three and married four times between the ages of twenty-two and eighty.[34] (If it is claimed that his married life was unhappy, there descends a new burden of defining marital happiness and proving it the norm during Russell's married years.)

Still, regardless of the conceptual problems with ahistorical psychological analysis, the fact remains that Descartes, Locke, Hobbes, Boyle, Newton, and Bacon (in spirit) were all bachelors. Marriage may not have been necessary for their professional success, and it may not have been desirable for other reasons. While a comprehensive discussion of their individual psychologies and emotional lives is beyond the present scope, it is also beside the primary interest in how their lives intersected with their work. This intersection has already been located in their professional success; it may be possible to pin it down further by examining the philosophical theories of the self that were developed during the seventeenth century.

If a philosophical theory of the self does not include emotional needs and wants, this does not mean that the self has no emotional dimension. The philosophical theory could be false or incomplete. But suppose a philosophical theory of the self that does not include emotional needs and wants is constructed in a historical context when people did not generally talk or behave as though they had emotional needs and wants. It would then seem that psychological theories of emotions that were developed in more emotionally "warm" cultural contexts could not be applied directly to the "cold" culture without anachronism.

It is difficult to see how anything more stringent than an emotional lament could be used to "evaluate" Boyle or Locke psychologically, for example. Are we to say that had they been "better adjusted" they would have been happier? Better adjusted to what? The idea of happiness as a secular value presupposes the importance of private feelings. Perhaps these men were more interested in their work than in how they felt. Their culture may not have provided a foundation for anyone to be importantly concerned with how he or she "felt." If that possibility is taken seriously, then once again there is no purchase for the psychological analysis, no way it get it going for subjects within that period.

Perhaps it was normal for men in the seventeenth century not to be concerned about feelings. Perhaps the bachelors of science were able to use their bachelorhood in life to strengthen what was most important to them about themselves, without experiencing any emotional conflict between their work and their personal lives. Thus far, I have been suggesting that their interest in the new science was most important to them among concerns in the world. In the next two chapters, through interpretation of two of Locke's philosophical theories of personal identity, I will show what was most important to the English bachelors of science as the *persons* they were. That is, the discussion will now move toward identity in the philosophical sense.

Five

Locke's Forensic Self

> [T]his meaning, this thought, which is in one sense perpetually active as if I were perpetually forming it, as if a mind were conceiving it without respite—a mind which would be *my* mind—this thought sustains itself alone by being; it by no means ceases to be active when I am not actually thinking it. I stand to it then in the double relation of the consciousness which conceives it and the consciousness which encounters it. It is precisely this double relation which I express by saying that it is *mine*.
>
> —JEAN-PAUL SARTRE, *Existential Psychoanalysis*

There is a richness to twentieth-century literary, psychological, and philosophical descriptions of the self that is easy to take for granted and assume was always present, if only implicitly, in Western life and writing. We overlook Jean-Paul Sartre's observation that much about the self has been created in the process of description. While it does not follow from this that when certain aspects of the self have not been described, they have not existed, the absence of description should signal the possibility of nonexistence. The term self, as a name for the introspector that encounters itself in introspection, first appeared in the English language only during the mid-seventeenth century;[1] and as we shall see, that early modern English sense of self was closely connected to what we now mean by the word conscience.

Locke is taken to have provided a theory of personal identity, and thereby of the self, in terms of memory. However, an alternative reading of Locke's enigmatic *Essay* chapter on this subject, together with an interpretation of that chapter by his eighteenth-century proponent Edmond Law, suggests that Locke did not have a theory of the self as we assume it. Instead, Locke's theory of personal identity was a combination of the religious confessional type of self first written by St. Augustine, and the forensic or legal-role self familiar in early modern English courts of law. There is no reason to believe that any of the other scientific thinkers in the seventeenth century had later, modern concepts of the self either—although, in an odd way, Descartes' "I," as a thinking substance identical with his soul, was more secularly modern than Locke's self because it could be fully experienced in the here and now. Locke's self seems to have been fully accessible only on Judgement Day, after death and resurrection.

In this chapter and the next, I examine Locke's concept of personal identity in terms of selves and owners. Neither aspect of Locke's theory of personal identity explicitly excludes strong emotions or close affective relationships with other selves. But the interpretation of Locke I offer here does render anachronistic the ascription of deeply affective subjectivity to Locke, as well as several traditional philosophical interpretations of Locke's theory. I argue that the Lockean person is no more than a fragment of what we would consider to be a whole person today. Although the Lockean self is capable of introspection, its introspection would not have the same subjects as ours. It can, in short, be read as a suitable self for a seventeenth-century bachelor of science who was fulfilled in his single state.

It seems safe to assume that Locke was theorizing his experience in his writings on personal identity, rather than creating a new form of experience. Social conditions that favored bachelorhood and other demographics suggesting low affect in the culture were in place at least until 1660—for example, the financial parameters of marriage, widespread abuse and neglect of children, absence of children from parents in infancy and adolescence, and high mortality rates within the family. Although Locke was born in 1632, and the *Essay* and *Second Treatise* were published in 1689, both works were many years in composition and there is no reason to believe that Locke was advancing descriptions of the self that were meant to account for or advance the cutting edge of social change.[2] However, the care Locke took to keep the identity of persons clear of theories of substance was an innovation, which was immediately pounced upon by his opponents (as will be seen in chapter 7).

Because the subject is philosophical personal identity, there is an unavoidable complexity in interpreting Locke's account. The philosophical problems he opened up remain unresolved—as Michael Ayers notes, twentieth-century philosophical discussions of personal identity have advanced only slightly past Locke's parameters.[3] Moreover, the philosophical concerns cut across traditions. For instance, Sartre could hardly be called an heir of English empiricism, but his location of ownership in conscious awareness is close to Locke's formulation.

Locke on Personal Identity

In his opening paragraph on *personal Identity,* Locke says we must first consider what "*Person*" stands for, and he immediately provides a synopsis of the analysis that will follow and then tells us that a person is "a thinking intelligent Being, that has reason and reflection, and can consider it self as it self, the same thinking thing in different times and places" (*Essay,* II, xxvii, 9).[4] He then specifies that the self is caused by consciousness, and he equates personal identity with the self-conscious sameness of persons: "For since consciousness always accompanies thinking, and 'tis that, that makes every one to be, what he calls *self;* and thereby distinguishes himself from all other thinking things, in this alone consists *personal Identity,* i.e. the sameness of a rational Being" (II, xxvii, 9).[5] Before these passages, Locke has already defined identity, separately, as follows: Existence itself is the primary principle of individuation; one thing can exist in different places but it can only have one beginning; more than one thing cannot have the same beginning at the same place. Whatever begins to exist is the same thing it is for the duration of its existence but not before or after that duration, which means that existence is continuous in time.[6]

Locke's analysis of what he calls "personal identity" is consistent with his general definition of identity, and for persons as well as things in general, the ontological aspects of identity seem to be neglected because in both cases the focus is on sameness. When Locke first says that a person is a thinking, intelligent, self-conscious being, ordinary pre-philosophical intuitions about persons are not disturbed. His claims that consciousness is aware of itself as consciousness and that consciousness is the cause of the self are philosophically sedate. But when he says "in this alone [i.e., consciousness] consists *personal Identity,* i.e. the sameness of a rational Being," it seems that by "personal identity" he no longer means—if he ever did—the cause or constituents of persons. And from that point on in the analysis, his criteria for persons become his criteria for selves: "And so far as this consciousness can be extended backwards to any past Action or Thought, so far reaches the Identity of that *Person;* it is the same *self* now it was then; and 'tis by the same *self* with this present one that now reflects on it, that that Action was done" (II, xxvii, 9).[7]

As Locke's analysis proceeds, the sameness of a person becomes the sameness of a consciousness that perceives itself to be the same consciousness over time:

> For it being the same consciousness that makes a Man be himself to himself, *personal Identity* depends on that only, whether it be annexed only to one individual Substance,or can be continued in a succession of several Substances. For as far as any intelligent Being can repeat the *Idea* of any past Action with the same consciousness it had of it at first, and with the same consciousness it has of any present Action; so far it is the same *personal self*. (II, xxvii, 10).[8]

Here, it is important to note that for Locke, although consciousness "makes the self," the cause of consciousness is unknown. Locke says that this cause is probably an immaterial substance, but that it cannot be known that it is not a material substance; and in general, according to Locke, substances themselves are unknowable.[9]

Problems with Locke's Account

Locke does not have a theory of personal identity for "full-bodied" human beings as we ordinarily experience ourselves prephilosophically—even though he often wrote and has been read as though that is precisely what he did have. Strictly speaking, Locke has told us this: A person is an immediate self-aware thinking thing; a self is the continuous totality of the person's self-aware thoughts at different times; a same person is a self. As Locke puts it toward the end of his analysis: "Where-ever a man finds what he calls *himself*, there I think another may say is the *same* Person" (II, xxvii, 26, second italics mine).[10]

In other words, the identity of a person is the sameness of a person, which is a self. Some outside distinctions in the meaning of 'identity' are necessary in order to pinpoint what Locke is setting down about personal identity, and what he is leaving out:

1. Each thing is the thing that it is: This is (uninformative) *A-is-A identity*.
2. Things that exist in time have A-is-A identity over time: Let's call this *t–t_1 identity,* or *sameness.*
3. Each thing of a certain kind is distinct from other things of that kind: This is *individuative identity*.

4. All things of a certain kind have some traits that make them members of that kind: Lets call this *constitutive identity.*
5. An immediate consciousness is self-aware: Let's call this self-awareness *reflexive identity.*

Putting these five distinctions to work in the context of Locke's analysis, a self is a $t–t_1$ immediate consciousness with reflexive identity. We do not know the constitutive identity of consciousness because what makes it is unknown and perhaps unknowable; and the individuative identity of consciousness is known only so far as consciousness is attached to particular human beings with physical bodies. The ability to individuate a thing would seem to be related to knowledge of its constitutive identity because we can't count things if we can't identify them, and we can't identify them if we don't have criteria for what makes them members of the kinds to which they belong. If a same person is a self, this means that a $t–t_1$ person is a $t–t_1$ consciousness with reflexive identity. The result is that, according to Locke, persons do not have constitutive identity and, because they are distinct from human beings, they do not have individuative identity either. However, for almost everyone, prephilosophically, persons are closely connected with human physical bodies and thereby have the same constitutive and individuative identities that those bodies do.

If Lockean persons are presumed to be distinct from physical bodies, then his analysis has the defect that it tells us less about persons than we already knew prephilosophically, when we thought they were human beings. In effect, Locke has collapsed the ontology of persons into the reflexiveness of selves over time. There is a weak and a strong defense against this criticism of Locke. The weak defense is to claim that a person is not an immediate consciousness but a full self. A same person would then be a $t–t_1$ immediate consciousness with reflexive identity over time periods longer than $t \ldots t_1$. While this may capture more of the complexities of memory, nothing is changed in principle because individuative and constitutive identities are still missing.

The strong defense against this criticism of Locke is that the word "same" in Locke's "same person," or the word "is" in "a same person is . . . " means 'constitutive identity'. That is, a person is made up by, or assigned to the class of persons on the basis of, $t–t_1$ consciousness. This defense is theoretical because it does not indicate how there could be criteria for either the constitutive or individuative identity of anyone other than a self with reflexive identity. No allowance is made for third-person individuative and constitutive identity. As an illustration of this problem, consider the following situation.

A human being, Jay, is aware of discontinuities in his self-consciousness and wonders whether he has multiple personalities or is "possessed" by another person. Suppose that a Lockean therapist, Kay, examines Jay in order to determine whether Jay is the same person. In that situation, Jay would be interested in whether he is the same person, with $t–t_1$ identity, individuative identity, and constitutive identity, according to objective criteria; Kay, however, would be interested in how criteria for self-identity apply to Jay. For Kay, the discontinuities in Jay's consciousness might be sufficient to establish that he is not always the same person but several persons, each of which is the same person to itself. But that would not satisfy Jay because he is looking for a test by which to determine whether the different chunks of his consciousness are objectively the same individual person. That is, Jay is interested in criteria for both individuative and constitutive identities. Kay explains to Jay that it is not known and cannot be known what the constitutive identity of consciousness is, and Jay accepts this. But when Kay tells Jay that he is as many persons as there are chunks of his consciousness, each of which thinks it is a self, Jay is puzzled and thinks that Kay has done no more than redescribe what he already knows and put a stamp of approval on it.

It is paradoxical that Jay, in the first person, has a third-person point of view on himself, whereas Kay, in the third person, takes first-person points of view on Jay. But the paradox merely underscores the relative unimportance of points of view for determining the individuative and constitutive identities of persons. From its own point of view, each of Jay's personalities might be a self, but Jay, speaking perhaps through one of these personalities, can rationally ask whether he is the same person that is, whether he is one person as opposed to a number of persons equal to the number of his discontinuous personalities or "selves." Kay cannot answer him to his satisfaction because she has no objective criteria for individuative and constitutive identity.

Part of the problem is that the constitutive identity of reflexive consciousness is unknown (and unknowable if it is a substance), and part of the problem is Locke's apparent failure to allow for the existence of persons in any terms other than their reflexive experience from a standpoint within that experience. A flat-out defense of what Kenneth Winkler has called the subjectivity of Locke's analysis in this regard would be to assert that in some important sense persons are selves to themselves, as Locke claims, and that Locke's analysis underscores the right of first persons—*I*'s—to constitute themselves based on what they remember having done. Winkler calls this the "authority" of the self to constitute itself.[11] The problem with Winkler's defense is that it is only an argument from authority and as such fails to address the problems raised by the absence of objective criteria for the identity of persons in Locke's analysis. In effect, Winkler's defense goes no further than the affirmation of Kay, the Lockean therapist in the above example.

However, this could be only an apparent failure that is due to the textual order of Locke's analysis, and not a failure at all if the last sections of the analysis are taken into account beforehand. How this is so will be considered in the following section, but first, a closer reading of several nuances, in the order in which they do occur in the text, supports the view that Locke did not intend to reduce objective persons to subjective experience and that it never even occurred to him that he could be read this way.

Before addressing the identity of persons, he considers the identity of trees, animals, and human beings as objective entities, and there is no indication that he does not intend to continue in the same mode with persons. Furthermore, his collapse of the constitutive identity of persons into reflexive identity is by no means explicit, because he hardly talks about persons at all in his analysis of personal identity, but about *same* persons. (He doesn't talk about men or souls either, but about same men and same souls.) Although his approaches to the question of what makes the same person resemble arguments about what persons are not, the arguments are literally about sameness, that is, t–t_1 identity. For example, he claims that the same persons need not be the same substances:

> For it being the same consciousness that makes a Man be himself to himself *personal Identity* depends on that only, whether it be annexed only to one individual Substance, or can be continued in a succession of several Substances. (*Essay,* II, xxvii, 10)[12]

He makes an analogous claim that the same persons are not the same bodies:

> Thus we see the *Substance,* whereof *personal self* consisted at one time, may be varied at another, without the change of personal *Identity:* There being no Question about the same Person, though the Limbs, which but now were a part of it, be cut off. (II, xxvii, 11)[13]

And he claims that the same human beings need not be the same persons:

> But if it be possible for the same Man to have distinct incommunicable consciousness at different times, it is past doubt the same Man would at different times make different persons. (*Essay,* II, xxvii, 20)[14]

While these examples support a claim that persons are not substances, bodies, or human beings, they are not that claim. Indeed, Locke does not even explicitly make the weaker claim that persons are not *necessarily* substances, bodies, or human beings.

Law's Key to Locke's Account

Reading Locke literally, we do not know what persons are until he supplies the information that 'person' is a "*Forensick Term* appropriating Actions and their merit; and so belongs only to intelligent Agents capable of a Law and Happiness and Misery" (*Essay,* II, xxvii, 26).[15]

Locke's announcement that he has been analyzing the identity of forensic persons, rather than "ordinary" persons, restores his personal identity analysis to something resembling an analysis of the objective identity of persons, whereby beings that may or may not seem to be persons upon first awareness of them can be individuated and sorted into the class of persons, or into some other class. Thus, the t–t_1 identity of consciousness that Locke has offered as an analysis of the identity of persons would need to be bracketed as something that persons can look for in other persons who have already been identified or individuated as persons before the analysis. This was Edmund Law's interpretation of Locke, and it provides a helpful framework within which to assess Lockean personal identity.

According to Law, writing in 1769, 'person' as a forensic term denotes a "quality or modification" in man that categorizes him as an accountable moral agent, subject to law and deserving of reward or punishment:

> When we apply it to any man, we do not treat of him absolutely, and in gross; but under a particular relation or precision: we do not comprehend or concern ourselves about the several inherent properties which accompany him in real existence, which go to the making up the whole complex notion of an active and intelligent being; but arbitrarily abstract one single quality or mode from all the rest, and view him under that distinct precision only which points out the idea above-mentioned, exclusive of every other idea that may belong to him in any other view, either as substance, quality or mode.[16]

This suggests that a forensic person is either an ordinary human being viewed under a special category, or some specific trait or disposition of an ordinary human being. Law argues further that 'person' in Locke's philosophic sense is an (invented) abstract term or general idea that is useful to society. This term or idea is applied to that rationality of an ordinary person "so far only, as it makes him capable of knowing what he does and suffers, and on what account, and thereby renders him amenable to justice for his behavior." For Law, then, 'person' refers to a *mixed mode* of human beings.[17] Extending Law's interpretation, 'person' would typically be predicated to convey something such as the following: "The man who now stands before the court is conscious of the former facts, and is therefore the proper object of punishment."

In a vein similar to Law's, Harold Noonan observes that scholars who view Locke as a conceptual pragmatist could interpret him to mean that the term 'person' is a conceptual response to normative requirements. According to Noonan, when Locke says, "In this *personal identity* is founded all the Right and Justice of Reward and Punishment" (II, xxvii, 18), he means that personal identity is sufficient as well as necessary for responsibility (or what Noonan calls "warranted accountability").[18]

Against this interpretation advanced by Law and extended by Noonan, Ayers interprets Locke to mean that a person is an unknown and perhaps changing substance of which consciousness is the organizing principle, analogously to the way in which life is the organizing principle of changing animal bodies.[19] The problem with Ayers's interpretation is that, according to Locke, such a person-type substantial substratum of consciousness would be just as unknowable as an immaterial soul-substance. Winkler proposes similarly that Lockean persons supervene on ordinary material substances (or "substance-stages"), although he acknowledges that his proposal cannot be attributed to Locke.[20] But aside from the lack of textual support for Ayers's interpretation and Winkler's proposal, the imputation of an unknowable person-type substance to Locke does not allow for an explanation of what the individuative and constitutive identities of persons are, because persons belong to the knowable, everyday world of ordinary experience and substances do not. Therefore, the individuative and constitutive philosophical identity requirements of an analysis of personal identity favor Law's interpretation of Locke, if the analysis is to preserve the sense of our ordinary, everyday experience of what we call "persons."

If Law's interpretation is correct, then Locke may have assumed that his readers knew that 'person' was a specialized normative, and even legal term, and that they would assume along with him that persons were ordinary human beings in special normative modes or, to use classical concepts (which Law relied upon, as Ayers and Noonan both note), that 'persons' were the *personae* of ordinary human beings, specifically the *roles* they played in courts.[21] On this interpretation, common sense would individuate persons based on the human beings whose modes and roles they were in particular situations. The same forensic person could be distinguished from the same human being—and of course Locke did make that distinction—but a forensic person would always have to be associated with one or more human beings in order to exist and have individuative and constitutive identity. Indeed, how else, if not through human agency, and at some point physical activity, could a forensic person do those acts, the resulting memories of which form t–t_1 consciousness?

Furthermore, Law's interpretation accounts for why Locke did not say that persons are not human beings but confined his arguments to the question of whether the same persons are necessarily the same human beings. That the same forensic persons are not necessarily the same human beings is compatible with their in fact "being" human beings most of the time, and even with the same forensic persons being the same human beings most of the time. Ontologically, forensic persons are predicates of human beings, but they would not be parts of the (nominal) essences of human beings according to Locke, because reincarnation is possible and resurrection is probable; and more importantly for our own present purposes, as will soon become clear, forensic persons are only the occasional roles of some human beings.

In arguing against the legal-role interpretation of Lockean persons, Ayers claims that forensic consciousness, or self-identity, is "natural" consciousness, and that forensic persons are ordinary human beings viewed under ethical categories.[22] But Ayers overlooks those differences between forensic and ordinary consciousness that prevent them from being the same thing: Forensic consciousness is closer to conscience than ordinary consciousness; forensic consciousness is more perfect than ordinary consciousness; forensic consciousness coheres through an identity form of *owning* that does not commonly exist in ordinary consciousness. These differences are shaped by the importance of punishment for forensic consciousness, which makes it both more concentrated and more restricted than ordinary consciousness.

According to the *Oxford English Dictionary,* the word *forensic* means "pertaining to, connected with or used in courts of law," and its root is the Latin word "forum," which has always meant a public situation.[23] By the same source, the seventeenth-century meaning of "consciousness" was "knowing with oneself or others."[24] Noonan points out that in the seventeenth century the word *conscience* had an overlapping meaning with the word *consciousness,* due to their common derivation from *consciring,* which combined a prefix meaning "with" and the Latin root *scio,* which meant "I know." For Locke, the word *consciousness* was often interchangeable with the word *conscience,* although conscience did not have its present connotation of "moral judgment."[25] Pulling these threads together with Locke's analysis of person as consciousness, 'forensic person' would mean "a consciousness that knows itself in a public and specifically legal situation." Thus, forensic persons are not silent self-knowers but self-knowers who are forced to account for themselves. And, according to Locke they are so forced because they are concerned with their own happiness, which is at stake in forensic situations:

> This personality extends it *self* beyond present Existence to what is past, only by consciousness, whereby it becomes concerned and accountable, owns and imputes to it *self* past Actions, just upon the same ground, and for the same reason that it does the present. All which is founded in a concern for Happiness the unavoidable concomitant of consciousness, that which is conscious of Pleasure and Pain, desiring, that that *self* that is conscious, should be happy. (*Essay,* II, xxvii, 26)[26]

Thus,the component of 'conscience' that today means 'moral judgment' is provided not by the forensic person himself, but by those others whose role it is to judge him in forensic situations. And for Locke, in a way often ignored by his defenders, including Law, forensic situations go far beyond secular courts of law:

> at the Great Day, when everyone shall *receive according to his doings, the secrets of all Hearts shall be laid open.* The Sentence shall be justified by the consciousness all Persons shall have, that they *themselves* in what Bodies soever they appear, or what Substances soever that consciousness adheres to, are the *same,* that committed those Actions, and deserve that Punishment for them. (*Essay,* II, xxvii, 26)[27]

Also, for Locke $t\text{–}t_1$ consciousness with reflexive identity is an important component of justice:

> But in the Great Day, wherein the Secrets of all Hearts shall be laid open, it may be reasonable to think, no one shall be made to answer for what he knows nothing of; but shall receive his Doom, his conscience [i.e., knowledge] accusing or excusing him. (*Essay,* II, xxvii, 22)[28]

Thus the strictures and formalities of forensic situations can be viewed as evocative of a special form of ordinary consciousness or as constructive of the forensic aspect of forensic consciousness. However, in contrast to forensic consciousness, ordinary consciousness includes private self-knowledge without judgment or penalty. That is, we do not as conscious beings at all times live our lives as defendants. Neither do those situations in which we might receive praise and reward occasion the kind of psychological intensity and public accountability that situations of punishment do: Good deeds can be performed cheerfully but bad deeds become worse if their perpetrators are light-hearted; the wishes of benefactors to remain anonymous are more often respected than are the wishes of criminals and sinners not to be identified as perpetrators of their transgressions or, once identified, to be kept obscure. This may be because there is a more strict standard of justice associated with punishment than reward. It is a greater injustice if an innocent person is punished than if someone who does not merit it is awarded a prize; it is worse if the guilty go unpunished than if the virtuous are not rewarded—conventional wisdom tells us that virtue is its own reward but not that vice is its own punishment.

The perfection of forensic consciousness rests on its continuousness and accuracy. There are no gaps in the $t\text{–}t_1$ identity of a forensic person; a forensic person remembers all (relevant) acts that he has done and presumably that only he has done. Locke appeals to an intuitive sense of God's fairness towards human beings as the reason for such perfection: "For supposing a Man be punish'd now, for what he had done in another Life, whereof he could be made to have no conciousness at all, what difference is there between that Punishment, and being created miserable?" (*Essay,* II, xxvii, 26).[29] For Locke, it seems to be this perfection of memory that makes possible the claim that if an action is done by a part of a person but not remembered as having been done by him, then it is not his action: "For whatsoever any Substance has thought or done, which I cannot recollect, but by my consciousness make my own Thought and Action, it will no more belong to me, whether a part of me thought or did it, than if it had been thought or done by any other immaterial being any where existing" (II, xxvii, 24).[30]

Problems with the Solution

Such a brittle standard of responsibility could preclude the virtue of expanding one's self by reflecting on those traits, such as alcoholism, that result in actions that one does not remember having done. However, the standard can be tempered if care is taken to pick out the initial act in a causal sequence that led to the wrongful act one does not remember having done, such as the alcoholic's first drink. But this tempered standard of responsibility still depends on the perfection of memory once memory has been pointed in the right direction: One is able to identify one's present self with the doer of the initial act in question and remember all of the details accompanying the act; or, alternatively, one can furnish an equally accurate alibi.

Ordinary memory does not have that kind of perfection. We sleep, daydream, are distracted, forget, and often cannot fill in the gaps. We not only live with such imperfections but are comfortable with them, joke about them, and display them as accessories to our skills and intelligence; the recognized expertise of philosophy professors is (I hope) not degraded if they cannot remember where they have parked their cars. It could be objected that such memory lapses are tolerated because they concern unimportant actions. But we do not trust our memories in weighty matters either, hence the reliance on notes and appointment books throughout professional life, and the widespread importance of written records generally. The objection from importance could be pressed on the ground that mentally competent people do not forget committing murder or other serious crimes. But the memories of serious crimes committed would be parts of forensic consciousness, so the objection collapses.

However, the most distinctive difference between forensic and ordinary self-consciousness rests on Locke's use of the term 'own' in the context of self-identity. Locke's analysis of personal identity is a consistent application of his earlier analysis of the general relation of identity: A person, as known through first-person introspection, is a continuous series of actions and states of consciousness, each of which has the same actor who is the introspector, and all of which are the (same) self of the introspector. Thus, according to Locke, when a self identifies itself through memory, past actions are identified as the actions of the present self and as parts of the total self. When the present self remembers a past action, it *owns* it:

> That with which the *consciousness* of this present thinking thing can join it self makes the same *Person,* and is one *self* with it, and with nothing else; and so attributes to it *self,* and owns all the Actions of that thing, as its own, as far as that consciousness reaches, and no farther, as everyone who reflects will perceive. (*Essay,* II, xxvii, 17)[31]

This kind of Lockean ownership functions as acknowledgment and rests on the same identity of a self with its parts that constitutes ownership of those parts by the self.[32] The idea that individuals own themselves was an accepted theme in the rights rhetoric of the middle classes who gained more political power in seventeenth-century England.[33] In the *Second Treatise* Locke used the term 'property' to mean "life, liberty and estate," as well as to mean only "estate" or what would today be considered possessions external to persons.[34] His argument that individual labor individuated the ownership of goods previously held in common in a state of nature, rests on an assumption that labor is owned by an individual because it is part of him. This idea of self-ownership in the *Second Treatise,* whereby self-ownership is the foundation for the property rights of individuals, will be examined more closely in chapter 6. Here it is important to realize that in the *Essay,* Locke uses self-ownership as a foundation for the responsibility of persons.

In contrast to Locke's forensic usage, in ordinary experience consciousness and memory do not involve the sense that *I* am what *I* have done in the past. When persons do identify themselves by their past actions, they are more likely to describe a present self in terms of the effects of those past actions than in terms of the actions themselves—for example, "I teach in a university" rather than "I got my Ph.D.," or (except poetically) "I have two children" rather than "I gave birth twice." Furthermore, present sorrows and failings are sometimes viewed as the effects of what persons have had done to them in the past, so Locke should require that what the self has suffered, as well as what it has done, be part of its identity. But there is no mention of such passive components of self-identity in Locke's analysis, and that omission is in keeping with his focus on responsibility as responsibility for punishable actions.

Thus, the kind of acknowledgment suggested by Locke's concept of self-ownership in the *Essay* resembles admissions of the form, "Yes I did ———," in response to accusatory questions of the form, "Did you do ———?"[35] But, more broadly than in situations where punishment is at stake, when we do identify ourselves by our past actions, those actions may include important achievements or acts of heroism, and not only actions which must be acknowledged in the sense of "owned up to."

Finally, it is as though the self of a forensic person were an ever-present totality according to Locke. Law realizes that this is not even a hypothetical situation for self-consciousness. Instead, Law settles for a capacity to remember parts of the self.[36] But such a capacity would not be sufficient for Locke's analysis, if he is interpreted to mean that the self literally is constituted by actual, public acts of acknowledgment or owning up. And on Judgement Day, at least, there would have to be such acts.

To sum up, Locke's 'self' is ultimately a religious entity, accountable for divine punishment and complete only as it will stand on Judgement Day. It is not actually a thing because of its dynamic dependence on memory for the sake of responsibility. This dependence freed the self from being an immaterial soul-substance, much to the outrage of Locke's theological opponents, such as Edward Stillingfleet.[37] (The epistemological importance of their dispute will be discussed in chapter 7.) As no more than a responsible self at the Resurrection, the Lockean person does not have everything it needs in theory to be a functioning entity in society. Although Locke says that persons are motivated by their own happiness, he clearly means the happiness that will ensue on Judgement Day—if all goes well. Nonetheless, Locke's self has the potential of functioning as a citizen in civil society, once we add to it his concept of ownership based on what he writes about property in the *Second Treatise*.

In principle, there is no reason why what we would call a personal and emotional life could not be added to the identity of a secularized Lockean self and owner. The result would be a broadening of self-identifying awareness to include feelings, motivated by a desire for happiness in this life. Again, there would be no underlying, unifying core to such a self. Neither would this self be accountable only in the contingency of punishment and blame because reward, praise, and even pleasure could be possible results of earlier choices made.

This reconstructed Lockean self would be the locus of an ordinary rather than a forensic person, distinguished from a Cartesian self by its insubstantiality. The connection of that self to a particular body would be no more than a strong empirical association. Freed in principle, but not in fact, from biology, this self could begin to assemble all of our agreed-upon general criteria for humanity. It could be the ethical subject to itself, as well as the ethical object to others, and it would be necessary to posit its existence for all members of an ethical domain, regardless of their biology as socially constructed in terms of gender or race.

That such a self reads like a philosophical invention, especially without a religious foundation, should not be surprising. It is a self based on mental experience rather than action or relations to others, and its identification with consciousness emphasizes its isolation. Even when reconstructed to include affect, with an appropriate neo-universalized moral worth, its existential flatness might still disappoint present ideals of self-fulfillment. Full introspection would be possible, but without an impetus toward the world and close ties to others the introspected domain would seem barren to us. If, as suggested earlier, Locke was offering a philosophical account of the type of person he knew best—namely, a seventeenth-century bachelor of science—then his theory of personal identity supports the historical intuition that the psyches of these men were radically different from their twentieth-century counterparts.

Six

Propriety and Civic Identity

To possess is to be united with the object possessed in the form of appropriation; to wish to possess is to wish to be united to an object in this relation. Thus the desire of a particular object is not the simple desire *of* this object; it is the desire to be united with the object in an internal relation, in the mode of constituting with it the unity "possessor-possessed." The desire *to have* is at bottom reducible to the desire to be related to a certain object in a certain *relation of being*.

—JEAN-PAUL SARTRE, *Existential Psychoanalysis*

The concept of property and property itself were of primary importance to Locke as a political theorist. However, more than the material possession of objects or even abstract rights was at stake in his attempts to justify individually owned private property, because property for Locke meant not only possessions but life and liberty as well.[1] Property for Locke is what a member of society *owns* that cannot be alienated without the owner's consent; and even consent cannot legitimately alienate life and liberty because they belong, not to the individual human being who is alive and free, but to God.[2] The connection between what is owned in this way and its owner reduces to a relation of identity. If something is a part of me I own it and I am the sum of my parts, each and all of which are me.[3] In Jean-Paul Sartre's terms, it could be said that for Locke *having* is a form of *being,* but the origin of being is not desire, as it was for Sartre, but God. The seventeenth-century combination of ownership, identity, and God-given rights was known as *propriety,* a concept that is almost opaque now due to later restrictions of things owned to material objects and the separation of 'rights' from 'ownership.' This concept of propriety was often connected with rhetoric against absolute monarchy. Pashal Larkin cites these words of an anonymous pamphleteer in 1644: "[Let us] who are English subjects . . . blesse God for His goodness who hath . . . *made us absolute proprietors of what we enjoy, so that our lives, liberties and estates,* do not depend upon, nor are subject to, the sole breath or arbitrary will of our Soveraigne."[4]

For Locke, humankind owns the world and is in turn owned by God.[5] Ownership is transitive because I own what my possessions own, such as "the grass my horse has bit" and the "turfs my servant has cut." Divinity is also transitive because man, as God's creature, is divine, and man's divinity spreads to what he owns—if it is not already directly imbued with God's divinity.[6] This transitivity of ownership and divinity, and of divinity through ownership, in principle put Locke close to Deism because it entails that all of nature and man's possessions are divine. But Locke did not appeal to Deism as a justifying vision. He intended to justify his political theory with established Protestant theology and natural law.

Overall, Locke's political theory in the *Two Treatises of Government* can be read as a rationalization of the economic and social order that would most benefit the class that Locke served in other ways as well. This was the class of parliamentarians and men of commerce, led by Locke's patron the earl of Shaftesbury and his Whig successors, who put William and Mary on the throne. Their interests lay somewhere between the exploitative market capitalism imputed to Locke by C. B. MacPherson and the just libertarianism that Robert Nozick derives from Locke.[7] (Locke was probably more religious than MacPherson credits and more biased than Nozick admits.) In any case, eighteenth-century political theory relied upon Locke's *Second Treatise* for the model that was used to justify Enlightenment human rights and the French and American revolutions. Robert Filmer's patriarchal model of monarchs and Thomas Hobbes' royal absolutism were contemporary contenders rendered defunct by the success of English parliamentarianism, although Hobbes' contractarianism has an analogue in Locke's location of ultimate political power in society before and after government.

My main task in this chapter is to examine how Locke's *Second Treatise* person, who is the individual unit in governed society, is an owner both of self and of other things. Again, as with Locke's Resurrection concept of a person in the *Essay* (the subject of chapter 5), what results is a "cold" person—emotions are not precluded but they would seem to be beside the point. In order to remain focused on identity, through the complexity of seventeenth-century political theory, my discussion will be limited to the following aspects of Locke's *Second Treatise* concept of ownership: propriety, individuality, property and labor, civic relationships, and solitariness.

Propriety

The dense cluster of ideas connoted by 'propriety' is evoked by the following often-quoted passage, written by Richard Overton in 1646, in a political tract titled, *An Arrow against all Tyrants and Tyranny shot from the prison of Newgate into the Prerogative Bowels of the Arbitrary House of Lords, and all other Usurpers and Tyrants whatsoever.*

> To every individuall in nature is given an individual property by nature, not to be invaded or usurped by any: for everyone as he is himselfe, so he hath a selfe propriety, else could he not be himselfe, and on this no second may presume to deprive any of, without manifest violation and affront to the very principles of nature and of the Rules of equity and justice between man and man; mine and thine cannot be, except this be: No man hath power over my rights and liberties and I over no man's; I may be but an individuall, enjoy myselfe and my selfe *propriety,* and may write my self no more than my selfe, or presume any further.[8]

This passage seems to amount to the following argument: Everyone who exists has a natural right to exist as that existing person, because otherwise he would no longer exist. This right ought not to be violated by others. The right is something that every person who exists *owns* and as such it is a possession or a property that ought not to be trespassed upon or taken over by others. This right to exist without interference is 'self-propriety'. Natural principles and basic human justice require that self-propriety not be violated: I may not violate the self-propriety of another and no one may violate mine. Unless self propriety is respected in this way, ownership of anything else is impossible.

Overton's implied claim that all ownership rests on individual ownership of an inviolable right to life must refer to all legitimate ownership, because a person can kill another and steal the other's property. But beyond the claim that legitimate ownership cannot be acquired by murder, it is not clear what Overton means in saying that there is no "mine" and "thine" without self-propriety. He seems to be asserting more than the obvious fact that dead people cannot own anything. He seems to be groping for a syntax of rights that, in its claims for natural individual rights, appeals to the already recognized importance of property. But what is particularly interesting to a present-day reader is that Overton does not yet fully use the language of rights, and he implies, through the use of the concept of propriety, that ownership is deeper and more important than a relationship between persons and material objects because it pertains to persons and their very existence.

Larkin remarks on the similarity between the Overton passage and Locke's chapter "Of Property" in the *Second Treatise,* and he relates Locke's broad sense of 'property', which includes life and liberty, to earlier seventeenth-century economic change and political thought and action. According to Larkin's historical account, the division between great landowners and businessmen was never sharp in England; on the continent also, in the sixteenth century, currency depreciation forced landlords to manage their estates more efficiently. Along with the increasing enclosure of arable land, fortunes made in trade and in the finance of trade resulted in a new propertied class. The sons of landowners became rich through trade; and the sons of tradesmen bought land. Seventeenth-century English landowners, who often were also merchants and investors, tended to profess Puritanism and embrace parliamentarianism. Their Puritanism required that they support religious toleration as a protection from the Church of England, and their parliamentarianism was intended to protect their growing financial enterprises through representation in the House of Commons.

This rising bourgeoisie had its political theorists during the years leading up to the Civil War and throughout the Restoration.[9] When members of the House of Commons introduced the Free Trade Bill of 1604, the right of every individual to trade freely and choose his own livelihood was presented as a natural right. James I and the House of Lords thwarted all legal recognition of such rights, but the free trade and commercial rights movement continued through activities in Commons until Charles I tried to govern without Parliament. Finally, desperate for revenues, Charles issued writs for new taxation in 1634–36, followed by further infringements on private property. The consequences of Charles's refusal to compromise on the issue of economic rights led to the Civil War.[10] Colonel Rainborow—Thomas Rainborough, whose festive funeral briefly solidified the Levellers' limited franchise movement in 1648—argued for universal suffrage in the choice of parliamentary delegates on the grounds that "the poorest hee that is in England hath a life to live as the greatest hee." But Henry Ireton objected with the claim that a natural right to suffrage implied a natural right to equal property. By 1657, during a debate on the reestablishment of the House of Lords, arguments for rights to political representation were openly based on amounts of land owned.[11]

Locke's use of 'property' to mean 'propriety' would have made perfect sense to his political colleagues, whether they read the *Second Treatise* in private circulation, as ideological encouragement during the events leading up to the Glorious Revolution, or consulted it after its 1689 publication, once the gains of the new bourgoisie had been legally institutionalized.[12] Locke's writing, unlike the rhetoric of his activist predecessors, sustains a somewhat rigorous interpretation of the connection between ownership and identity, to which I will now turn.

Individuality

In Chapter V, "Of Property," in the *Second Treatise,* Locke begins by setting himself the task of showing how property can be possible without everyone's consent, when God originally gave all of the earth to all human beings in common.[13] It seems clear that Locke is here already assuming that property cannot be taken from its owner(s) without consent. This means that he takes for granted beforehand the existence of property rights, and that his task is to show how individuals come to own part of the whole of the earth when originally everyone owned everything in common. That is, he is looking for a derivation of the right to *private* property based on natural law, or the law of God given to man in a state of nature.[14] He also assumes that we know the following from either reason or revelation: "Men, being once born, have a right to their Preservation, and consequently to Meat and Drink, and such other things, as Nature affords for their Subsistence" (II, v, 25).[15]

Locke further took it for granted that nourishment and other necessities for survival need be consumed by only an individual for that individual to survive. Therefore, in deriving private property rights, Locke assumed that such private property was in a fundamental sense owned individually rather than collectively. If he had realized that some goods need to be consumed collectively for survival, such as those goods that are consumed when, according to Locke's own analysis of the family (discussed in the final section of this chapter), parents fulfill their God-decreed obligations to take care of their children, he could have posited the family as well as the individual as an owner of property. Who or what is an owner is very important at this juncture of Locke's political theory because owners are the basic units, persons, or citizens in political relationships. If Locke had posited the family as an owner, then the family would have been the unit person in political society.

Linda Nicholson and other recent feminist scholars have analyzed

Locke's *First Treatise* and *Second Treatise* arguments against Robert Filmer as a reification of the separation between the family and the state that was a contingent effect of the historical circumstances in which Locke wrote.[16] This analysis is supported by Locke's strong emphasis on the ways in which family relationships could not be used as a model for political relationships because family relationships are temporary and parents do not have life and death power over their children.[17] Locke's insistence on asymmetrical parental duties to children inadvertently blocked any broad arguments for a welfare state, as did his suggestions that poor children earn their keep by the age of three, so as not to be a burden on their local parishes.[18]

Nicholson calls the Lockean separation of family from government a reification because it projects or generalizes the seventeenth-century transition from extended kinship systems to nuclear families onto the state of nature and presents this as a universal condition of human life.[19] However, there was an equally important reification of the individual in Locke's derivation of private property, which established the individual and not the family as the unit person in society. The historical contingency of this second reification was the ability of people such as Locke and the other bachelors of science in the seventeenth century to survive and thrive without families.

Locke was aware that some property in the family was common property. And, as Nicholson points out, he stipulated that some property in the family was owned by wives independently of their husbands.[20] But Locke derived private property from a God-given natural right to survive, and it seems simply not to have occurred to him that in an actual, historical state of nature, individuals could not have survived without families. Therefore, common property within families, or property belonging to wives, would not have seemed important enough to Locke for him to posit an entire family as a collective owner-person in society.

Property and Labor

In deriving individual private property rights, Locke relied upon a concept of identity whereby the individual could be assumed to own what he already is or what is already a part of him. As a result, ownership based on identity came to support property rights, in a mirror analogue of the way identity based on ownership came to support responsibility in the *Essay* chapter on personal identity: "The Fruit or Venison, which nourishes the wild Indian, who knows no Inclosure, and is still a Tenant in common, must be his, and so his—*i.e.* a part of him, that another can no longer have any right to it, before it can do him any good for the support of his Life" (II, v, 26).[21] This passage makes it clear that for Locke it was in some sense analytic that if something is a part of an individual, then that individual owns it.[22]

Presumably, an individual is the sum of his parts, so Locke's justification of ownership rests on an implicit assumption about identity: I am the sum of my parts; I own each of those parts; I have a right to live; I can't live without my parts. Therefore, my ownership of my parts is protected by my right to live. If anyone removes a part of my body or otherwise harms me, then that person has not only stolen my property but violated my right to live.

In this assumption that identity implies ownership, it is not clear whether the concept of property is being used to protect the concept of a right to live, or whether the concept of a right to live is being used to justify the concept of property. The correct reading is probably that Locke in the *Second Treatise* is attempting to construct a model of civil society where the concept of property can do all of the work necessary to safeguard what he must have intuitively understood to be basic human rights; like Overton, however, he does not yet have a complete language of rights.

But Locke did have a complete language of property. He was interested in more than just a right to goods necessary for survival—he was interested in property as an institution in complicated, civilized society. He therefore extended the ownership of parts of an individual's body to the ownership of an individual's labor by that individual. Locke uses labor to justify ownership, not because we deserve to own what we have worked on—which he does not insist upon—but because an individual owns his labor. Thus, Locke's derivation of private property from labor is more metaphysical than moral, because

> every man has a *Property* in his own *Person*. This no Body has any Right to but himself. The *Labour* of his Body, and the *Work* of his Hands, we may say, are properly his. Whatsoever then he removes out of the State that Nature hath provided, and left it in, he hath mixed his *Labour* with, and joyned to it something that is his own, and thereby makes it his Property. (II, v, 27)[23]

Therefore, according to Locke, who already assumed that we know what ownership is, an individual owns his body and thereby owns his labor and whatever his labor has been 'mixed' with. Ownership has to be dynamic for this analysis to go through, though the nature of its movement is unclear. It is tempting to say that ownership is transitive according to Locke. But if *A*, in owning *B*, owns *C*, which is owned by *B*, it is not clear precisely how the results of labor are "owned" by labor, which is an activity rather than a person, so that these results of labor could belong to the owner of that labor.

However, there is an even more serious problem if Locke's dynamic analysis of ownership through labor, is allowed to go through: How do I come to own the parts of something that were untouched by my labor so that I own the whole of something when my labor has only come in contact with part of it? Robert Nozick brings this question to absurdity when he asks whether I own the whole of the sea after mixing my can of tomato juice into it, or have merely "foolishly dissipated" my tomato juice.[24] Less whimsically, suppose that an individual or a nation owns a parcel of land, supported by positive law and possession. How does the owner come to own the mineral rights of that land, and how far down does that ownership extend and why?

If this Lockean alchemy of the extension of ownership through individual labor cannot be satisfactorily explained, then the more traditional problems with a labor justification theory of property arise. Presumably, labor justifies property because people deserve to own the fruits of their labor. But many labor all their lives and own little or nothing, and many owners have never labored. Also, communal labor, such as the construction of a house or all of the agricultural and manufacturing contributions to the end product of a loaf of bread, result in no clearly specified ownership by individual laborers.[25] The result is either that labor justifies ownership only in certain specific cultural contexts, such as the labor of an industrialist, or that labor is worth pursuing for its own sake.[26] In the first case, the determination of ownership is unrelated to labor; in the second, the necessity of labor has been eclipsed by its moral value. I find the second result more interesting historically because it points to the divine nature of labor in Protestant ideology.

Civic Relationships

According to Locke, civic relationships are based on the importance of property. Individual owners were able to survive peacefully and productively in a state of nature, given two provisos and a fundamental law of nature. The provisos limited appropriation so that there would be no waste or spoilage of goods appropriated and any act of appropriation would leave "enough and as good" left over for appropriation by others.[27] The law of nature was that individuals may not harm one another. Where there was harm, which would include theft, the injured party in a state of nature had the right to punish the offender and seek redress for the loss.[28] These rights were often inconvenient to enforce, so individuals agreed to give them up to a government that would justly enforce them for individual citizens. Thus, human society existed prior to government, and the sole purpose of government or civil society was to protect the ownership of the life, liberty, and property of the members of that society.[29]

On Locke's account, all of the complexities of government could be reduced to the model of a contractual agreement for the protection of property. Strictly speaking, so long as life and liberty were part of property, there would be no need for a concept of abstract rights, as they came to be formulated during the Enlightenment. This model could be reclaimed today if life and liberty were constructed as goods owned, although there is a problem with providing an explication of the meaning of ownership that does not refer back to life and liberty and therefore to rights in some sense. It might work to insist that human identity consists of life and liberty and that, as far as individual persons are concerned, their property in themselves is inviolable. But then the model breaks down when it is extended to property in the sense of material possessions. My car and computer, as important as they may be, are not me.

It is less than elevating that in the seventeenth century the concept of property, in the narrow sense of ownership of material possessions, was used as a metaphor for the ownership of more important nonmaterial goods, and thereby as the basic end or purpose of political relationships. Conal Condren, in *George Lawson's Politica and the English Revolution,* points out that there was a strong tradition behind this metaphor, beginning in fourteenth-century Italian associations of the status of rulers with the absolute rights of owners.[30] Since the rhetoric of ownership was originally used to support the power of kings, by locating strong property rights in individuals both Overton and Locke can be interpreted as having stood this tradition on its head—they relocated the absolute rights of royalty in individual citizens, through what reads today as harsh materialism. The absolute rule of monarchs had been harsh, and all classes in the seventeenth century presumably understood this historical harshness of the royal possession of all land and subjects. Locke's relocation of those absolute, royal rights in individuals must have been effective ideological prose for his time and, not incidentally, would have strengthened the identity of citizens as they saw themselves. The rights of kings were transformed into the rights of individuals through the rhetoric of Overton and Locke. This rhetorical shift was liberating for the early modern bourgeoisie because government was oppressive to them. Locke's distinction between society and government, which legitimizes revolution if citizen's rights are violated, would work differently if society were malevolent and oppressive and government benevolent and liberatory. In that case, the location of absolute property rights in society, independently of government, could be used as a justification for revolution against a government that was otherwise just.

Solitariness

It is frequently assumed, in both traditional and feminist political theory, that after the seventeenth century a legal person was a male head of a nuclear-family household. It is true that until the twentieth century women had few rights in law and married women could not vote, own property, sign contracts, or obtain child custody after divorce. However, Locke's analysis of property and the family does not include familial or other personal ties in the identity of a citizen-owner. Parents are obligated to take care of their children but this obligation ceases when the children are grown.[31] Children in turn enter into their own relationships with society and government as soon as they accept an inheritance of property or otherwise acquire property legally; they do not inherit their parents' citizenship.[32] Adults are expected to respect their parents and be friends to them if they were well-treated as children, but they have no obligation to obey them as they did when they were minors, and no obligation to pay them back for care or goods received.[33]

Locke's account of marriage is as atomistic as his account of the parent-child relationship. Men remain with women after children have been conceived because human infants require the care of both parents. The conjugal tie is prolonged because women are able to conceive again before young children are grown. Married couples therefore remain together in order to raise their children, and it is assumed that the main personal reason for marriage is procreation.[34] Even though mutual affection may develop between spouses, there is no reason why they cannot go their separate ways once the parenting task is complete.[35]

Thus, Locke's analysis of marriage and the family adds nothing to the connection between a citizen-owner and others. Before the law, any rights that married women might otherwise have were referred back to the rights of their husbands. When Thomas Blackstone codified English law in the eighteenth century, married women had the legal *coverture* of their husbands. But the concept of a person as a male head of household was more a result of how women were treated in law than it was a direct component of personal identity insofar as men were citizen-owners. When Locke argues for the rights of individuals to appropriate property and enter into agreements for the protection of their ownership of that property, he seems content that this is the major issue in citizen-government relations. He does not, for example, speak of any natural right to enjoy life as the member or head of a family. He does not situate his owner-laborer in the state of nature as someone whose fundamental well-being is in any way connected with personal relationships. Neither does he mention the acquisition of knowledge as a natural right. Of course, rights to learning and publication, as well as procreation, could be derived from property rights insofar as those include life and liberty. But the point is that Locke's main concern was with ownership, and if one constructs identity as the sum of the parts of what one is, each of which one owns, then the basic unit of society is one person. This person would be a self-aware consciousness, accountable to God for everlasting happiness, and all other persons would be accountable to him or to the government, on his behalf, for any infringement of his propriety. In theory, according to Locke, nothing more would be required for the identity of a human being. In fact, he seems to have believed that nothing more than this could be expected from life, for he wrote in a letter to a friend: "This world is a scene of vanity . . . and affords no solid satisfaction but the consciousness of doing well, and the Hopes of another Life. This is what I can say by experience, and what you will find, when you come to make up your account."[36]

Apart from the "enjoyment of possession," which was a common locution in seventeenth-century political ideology, one wonders what might have relieved the isolation at the core of any bachelor of science. One relief would have been the collegiality 'enjoyed' through the shared enterprise of the new science. Further consolation, if it were necessary, could be found in the virtuous nature of personal isolation. Secrecy and suspiciousness were traits to be cultivated among the propertied and intellectual elite. For example, Lawrence Stone quotes Sir William Wentworth's 1607 "Advice" to his son Thomas, who later became earl of Stafford:

> Be very careful to govern your tongue, and never speak in open places all you think. . . . But to your wife, if she can keep council (as few women can) or to a private faithful friend, or some old servant that hath all his living and credit under you, you may be more open. . . . Ever feareth the worst. . . . Whosoever comes to speak with you comes premeditate for his own advantage.[37]

Stone describes this attitude as an early seventeenth century "numbing of affect." But there is no evidence that there was any relevant underlying affect to be numbed.

Bacon's affairs were always complicated and he openly advised cultivating the trait of secretiveness: "The best composition and temperature is to have openness in fame and opinion; secrecy in habit; dissimulation in seasonable use; and a power to feign, if there be no remedy."[38] There is a pleased matter-of-factness about Bacon's advice on these matters which suggests that he did not find the subject of dishonesty painful or shameful.[39] Descartes lived most of his adult life in Holland, where he moved more than thirty-five times, often without informing his friends, and he enjoyed confusing other mathematicians.[40] Hobbes took a long time in answering direct questions.[41] Locke, Boyle, and Newton all used codes to keep records or write letters about controversial political or religious matters, delicate personal matters, and, in the case of Boyle, alchemical formulae.[42] Locke addressed secrecy very directly in his *Copy Book,* in a way reminiscent of both Wentworth and Bacon: "Tell not your business or design to one that you are not sure will help it forward. All that are not for you count against you, for so they generally prove, either through folly, envy, malice, or interest."[43]

It was undoubtedly prudent to cultivate secrecy, and it seems to have been a mark of wisdom and status to advise others how to do so. The pleasure taken in this rhetoric of what we would today call alienation, or even paranoia, must have been a compensation for isolation—if the personal isolation of the bachelors of science were otherwise cause for unhappiness, which we do not know. Even the word "unhappiness," when cast back in this way, could be anachronistic because there seems to have been no widespread social ideal of personal companionship or romantic or familial love that today would be considered necessary ingredients of individual happiness.

The disintegration of traditional family connections and struggles in work environments in the late twentieth century cause forms of disconnectedness and alienation that may be similar to the seventeenth-century chill over personal relationships. When we suspend our own expectations for happiness and rely on our selves in circumstances that normalize detachment and "paranoia," the seventeenth-century (stoic) location of wisdom in discretion may be relevant.

Even now, not everyone has the type of affective self that gives rise to disappointed expectations. Those who do might be appalled by the Lockean model and point out that my apparent suggestion that it be emulated contradicts the suggestion at the end of the previous chapter that the Lockean self could be "warmed up." However, the point of both suggestions is that the emotional dimensions of selves are variable components of human identity that depend unreflectively on cultural consenses and reflectively on evaluations of these foundations. The cultural nature of emotional expectations—and derivatively, the cultural hook on emotional experiences—means that choices can be made and new theories constructed about what previously seemed to be uncontrollable because it was "natural."

Seven

Protestant Difference and Toleration

> This only I say, that however clearly we may think this or the other Doctrine to be deduced from Scripture, we ought not therefore to impose it upon others, as a necessary Article of Faith, because we believe it to be agreeable to the Rule of Faith; unless we would be content also that other Doctrines should be imposed upon us in the same manner.
>
> —JOHN LOCKE, *A Letter Concerning Toleration*

In his "Letter to the Reader" in the *Essay,* Locke presents the work as the result of an investigation occasioned by difficulties in a conversation among a group of friends who met in his chamber.[1] Insofar as the *Essay* was supportive of natural philosophy or empirical investigation, it could and did furnish the philosophy of science for members of the Royal Society, apart from Locke's skepticism about the certainty or perfection of scientific knowledge.[2] But philosophy was not itself an institution, and the ongoing investigations by active members of the Royal Society required institutional support, funding, and approval. Oxford and Cambridge were still too steeped in scholasticism and theology to provide this function in the seventeenth century, and almost as a creed Bacon, Locke, Hobbes, and Boyle repudiated their curricula. Newton was Lucasian Professor of Mathematics at Cambridge University, but his scientific work was not directly supported by that post.[3]

One wonders why religion could not have furnished institutional support for scientific investigation in seventeenth-century England. Why didn't Protestantism support the advancement of knowledge, as Catholicism had purported to do for centuries in Europe? Thae practical answer is that

Protestantism had not been in existence long enough; the intellectual answer is that even though all of the bachelors of science seem to have been personally religious and religion was incorporated into educated views of human nature and political theory, English Protestants were not united within one church, as Catholics traditionally were. Indeed, the quote from Locke's *Letter on Toleration* at the beginning of this chapter epitomizes the liberal Protestant view that, given incompatible creeds, pluralistic coexistence was the only reasonable solution.

Protestant theological disputes were part of life-and-death politics in the first half of the seventeenth century, which made it unlikely that scholarly or scientific work could have progressed under the the umbrella of any one religious institution. However, between 1650 and the publication of Locke's *Essay,* a group of Anglican (Church of England) divines known as Latitudinarians furthered toleration among all Protestants and advanced a view of probable truth, known as *moral certainty,* that was coincident with scientific evaluations of empirical knowledge among members of the Royal Society. Eventually, the controversy between Locke and Edward Stillingfleet over the identity of persons and knowledge of substance put an end to this epistemological coincidence. Locke's empiricism was the dominant influence in English philosophy after that dispute, and his pragmatic concept of toleration carried on the spirit of Latitudinarian toleration. The knowledge claims of Newtonian science exceeded Latitudinarian moral certainty, and by the end of the century the Newtonian paradigm became the cohesive force within the Royal Society. During this early mercantile period, all serious enterprise had government protection through the crown, Parliament, or both. Therefore, given the need for institutional support, it was just as well for science that the Royal Society had government sponsorship from its inception—even though royal funding was more readily promised than delivered.

In this chapter, I first briefly examine the political context of several differences among English Protestants. I then move to the scientific epistemology of the Latitudinarians and the point at which science and empirical philosophy parted ways with that religious liberalism. Finally, I compare the practicality of the religious toleration exemplified by Locke with John Milton's more idealistic concept of toleration. With this background in place, it will be possible in chapter 8 to return more directly to the subject of identity, through an independent look at the epistemic and character ideals set forth by scientists and virtuosi of the Royal Society.

Puritanism and Politics

Disputes among English Protestants became politicized during the reign of Charles I. When Charles's archbishop, William Laud, attempted to impose the Anglican Prayer Book on Presbyterians in Scotland, rebellion resulted and the king was forced to summon Parliament in 1640. Laud was imprisoned and the king's minister, Thomas Wentworth, was executed. At that time, which was the beginning of the Long Parliament, the Puritans were the majority in Parliament, but they factionalized during the Civil War years of 1642–46. It is important to remember that English religious disputes all took place within Protestantism. Henry VIII had established a Protestant Church of England in defiance of the Roman Catholic Pope in the sixteenth century, but English Protestantism itself divided into, on one side, the official crown's religion of Anglicanism, and on the other side, the Puritans, who, agreed that Elizabeth I had not completely reformed the Church of England, themselves divided into a multiplicity of sects.

Richard Dunn, after contemporary scholarship in which historians inconclusively assign preeminence to economic and political causes, argues for a religious interpretation of disputes during the Civil War period. According to Dunn, there were three main religious factions within the Puritan Long Parliament: Presbyterians who wanted the Church of England changed to a national Calvinist Church; Independents who wanted no national church at all and advocated religious tolerance among all sects; and radical sects such as the apocalyptic Fifth Monarchy Men, Levellers who argued for universal male suffrage, and Diggers who were against private property on religious grounds. Cromwell was an Independent, and the Rump Parliament he created in 1648 by excluding the Presbyterians was Independent as well. Although Cromwell extended religious toleration to almost all Puritan sects, he did not extend it to Anglicans and Roman Catholics (and his army dealt more harshly with Catholics in the annexation of Ireland, than it had with royalist Anglicans in the English war).[4]

When Charles II was restored to the English throne in 1660, he was financially dependent on Parliament. The Anglican Church became the official Church of England, but almost all of the Puritans, including Presbyterians, Independents, Baptist and Quaker nonconformists, were tolerated. Mercantilism began to shape political and economic structure when the Navigation Acts ensured the protection and expansion of English foreign trade by excluding Dutch middlemen. Colonial emigration increased, profits from American agricultural imports increased, and English merchants came to dominate the slave trade. At the same time, Charles enjoyed the support of an Anglican Cavalier (or royalist Tory) Parliament and made secret plans with Louis XIV of France to restore Roman Catholicism in England. In 1672, Charles tried to suspend Parliament's Clarendon Code by granting toleration to Catholics and nonconformists, but widespread public indignation forced him to retrench. During the 1670s the Tories, in allegiance to the crown, supported absolute monarchy and the Anglican Church of England. The Whigs, led by the earl of Shaftesbury (Locke's patron), were in favor of parliamentary supremacy and wider toleration of nonconformists. Both Tories and Whigs were anti-Catholic and anti-French. Both Tories and Whigs also belonged to the official Anglican Church of England, so their main doctrinal differences came down to the issue of toleration in regard to nonconformists and the strength of Parliament in relation to the crown.

James II, Charles II's brother, was openly Catholic, though his daughter Mary was married to William of Orange, heir of the ruling Calvinist Dutch family. After James's queen gave birth to a son, which threatened a Catholic succession, Whigs and Tories, in order to ensure a Protestant succession, united in 1688 to invite William to bring a foreign army to England. James fled and William became king without serious opposition. The Toleration Act of 1689 allowed Protestant nonconformists to worship publicly. Catholics were excluded, although their persecution thereafter was minimal within England. Whigs and Tories coexisted peacefully after the Glorious Revolution. This lasting peace was initially due to the effective representation of merchants and landowners by members of both parties in Parliament. From the 1690s on, Parliament and the crown were responsive to the interests of merchants and landowners and the costs of government were financed by taxation and public borrowing. Political peace and economic expansion was accompanied by an overall tempering of religious disputes.[5]

The Latitudinarian Solution

The road to the English religious peace at the end of the seventeenth century was paved by the theological, political, and even scientific efforts of the Latitudinarians—John Wilkins, John Tillotson, Edward Stillingfleet, Simon Patrick, Thomas Tenison, William Lloyd, and Gilbert Burnet, the colleague who chronicled the movement.[6] Recent scholars differ on the application of the term Latitudinarians. Barbara Shapiro uses the term 'latitudinarian' to apply to all who shared William Chillingworth's concept of moral certainty about probable truth, as an English epistemological *via media* between complete skepticism and dogmatism in religious and scientific knowledge.[7] Martin Griffin, in a posthumously published book updated by Richard Popkin in 1992, distinguishes the Latitudinarians from three other schools of thought: Lucius Carey Vicount Falkland's Great Tew Circle near Oxford during the 1630s, where Chillingworth's epistemic claims originated; the Cambridge Platonists who advocated moderation in religious disputes during the 1650s; and philosophical and scientific writers, such as Robert Boyle, who shared Latitudinarian beliefs about toleration.[8] I will conform to Griffin's more narrow application because I think it identifies the conceptual points on which philosophical empiricism and liberal Protestantism diverged.

Some writers trace the beliefs of the Latititudinarians to Socinian natural religion. Socinianism began with Michael Servetus, who first wrote a criticism of the doctrine of the Trinity in 1531; published *Chritianisme Restitutio* in 1552, which advocated a reform of Christianity through a return to its origins; and was burnt at the stake in Geneva in 1553. A school of Italian intellectuals continued Servetus's tradition until Faustus Socinus wrote in favor of universal toleration and separation of church and state in his contribution to the anti-Trinitarian movement in Poland toward the end of the sixteenth century. By that time, biblicism, the importance of reason in religion, and a new doctrine of the divine function (but not the divine nature) of Christ were added to the developing heresy. All of these ideas were received sympathetically in Holland, from where they traveled to England. The first Socinian theologians who came to England in the late sixteenth and early seventeenth centuries were themselves burnt and imprisoned, and explicitly Socinian books were burnt by public hangman throughout the seventeenth century. It is thereby understandable that few theologians or philosophers would openly admit to reading or possessing Socinian works, much less accepting Socinianism as the foundation of their beliefs.[9] A strong case can be made that Locke and Newton, for example, not only read Socinian theology but repudiated the official Church of England doctrine of the Trinity, even though they always professed devout adherence to Anglican principles. But in general, the complexity of English theological debate and its expression in real politics make it difficult to determine who was in fact a Socinian. The heretical status of Socinianism made it a potent term of abuse, and some writers may have been called Socinians in controversies that concerned matters other than theological disputes. (Also, since theological history is still written by theologians, how much Socinianism is read into seventeenth-century texts depends on the scholar's own leanings.)[10]

However, the ideas of the Latitudinarians, even in Griffin's narrow sense of the term, did have much in common with the teachings of Socinus. Although no one publically repudiated the doctrine of the Trinity and all were official Anglican divines, the Latitudinarians advocated a reasonable approach to Protestant Christianity, which amounted to a consensual core of the following claims: Religious truth cannot be known with demonstrative certainty, but only with moral certainty or conviction based on a high degree of probability. It is not necessary that all Christians perform the same ceremonies or share beliefs on all issues—there can be widespread differences on theological issues, so long as there is agreement on essentials. Reason can and should be used to interpret scripture, which is the ultimate authority for religious truth. No church is infallible, and individual conscience is the judge of interpretations of scripture. What goes on internally in individual conscience is a private matter, so outward signs of goodness are sufficient. Moral goodness is more important than skill in doctrinal analysis. Passionate excess or "enthusiasm" is to be avoided as much as dogmatic certainty. And finally, Christ is divine.

The Latitudinarian concentration on essentials and its relegation of religious conscience to a private subjective domain made possible toleration over a broad spectrum of Protestant beliefs. The acceptance of the use of natural reason to make probable judgments was analogous to the acceptance of empirical belief within science, and, as noted, many Latitudinarians were members of the Royal Society (and vice versa). Furthermore, the importance of private conscience would have been underscored by Locke's theory of personal identity, as discussed in chapter 5, whereby a person is an ultimately responsible religious consciousness. However, it is important to appreciate the division between theological Latitudinarianism, with its probable basis of belief, and empirical philosophical epistemology. Both systems were positions between dogmatism and skepticism, but thorough going philosophical empiricism eventually led to religious skepticism and a completely secular scientific tradition (apart from the personal religious beliefs of scientists).[11]

Stillingfleet versus Locke

As a Latitudinarian, Edward Stillingfleet, bishop of Worcester, was dogmatic about his religious reasonableness. For example, although he repudiated Catholic doctrines of the infallibility of the Church and of the right of religious officials to inquire into the inward state of an individual's conscience, he insisted that belief in the divinity of Christ be based not on faith but on reason, because the truth of scripture was highly probable.[12] Since, assuming monotheism, the divinity of Christ implied the doctrine of the Trinity, it was also important to Stillingfleet that the doctrine of the Trinity be accessible to rational understanding. Stillingfleet wrote voluminously on many subjects, including law and antiquity, and was believed to be more committed to his scholarship than his parochial duties.[13] However, Stillingfleet was not philosophically skillful enough to argue his Latitudinarian position successfully against Locke's philosophical empiricism and theory of personal identity. At critical junctures in their exchange, where he should have shown precisely how Locke's empiricism excluded the doctrine of the Trinity, Stillingfleet tended simply to call Locke a Socinian. Locke was sarcastic about Stillingfleet's ineptitude in a letter to his cousin, Peter King: "The Bishop is to prove that my book has something in it that is inconsistent with the doctrine of the Trinity, and all that upon examination he does, is to ask me, whether I believe the doctrine of the Trinity as it has been received in the Christian Church?—a worthy proof!"[14] Stillingfleet's charge that Locke was a Socinian was based on the use made of Locke's *Essay* by John Toland in his *Christianity Not Mysterious*. Toland was persecuted for his explicitly Socinian attempt to reduce all Christian belief to empirical principles. Locke had helped Toland personally, though he probably did not agree with his interpretation of his work and certainly did not appreciate having his association with Toland made public.[15]

Locke's positions on revelation and tolerance suggest that he may have been even more empirical about religion than Toland was. Locke may have thought that Christianity could not be made "not mysterious" and that where it conflicted with philosophical empiricism, the latter ought to prevail. But, given the importance of religion in the historical context of his dispute with Stillingfleet, it was out of the question for Locke directly to challenge the Latitudinarian assumption that Christianity could have an empirical foundation, and impossible for Stillingfleet intellectually to consider such a challenge. Locke could not have risked either explicit fideism or atheism, and Stillingfleet, as a minister of the Church of England and a Latitudinarian, could not have taken either position seriously. Nevertheless, the real issue of contention between them was whether reason and probable knowledge could support the Anglican doctrine of the Church of England. In terms of the history of philosophical problems, Stillingfleet's attack on Locke is inconsequential because it was theological rather than philosophical. But in the history of philosophy as a discipline, Stillingfleet's attack marked the beginning of the secularization of English empiricism and is therefore enormously important.

Locke professed himself content that a doctrine such as the Trinity be accepted on faith, a position that Stillingfleet opposed because he thought that this doctrine did not conflict with rational criteria for knowledge.[16] Early on in his reply to Stillingfleet, Locke argued for a distinction between faith and knowledge on the grounds that such a distinction was necessary to preserve the nature of faith: "Faith stands by itself, and upon grounds of its own, nor can be removed from them and placed on grounds of knowledge. their grounds are so far from being the same or having anything in common, but when it is brought to certainty faith is destroyed; it is knowledge then and faith no longer."[17] So on this point at least, Locke proved Stillingfleet's worst accusation by clearly professing an adherence to what was a heretical, if not a Socinian, claim—though of course he did not present it as such.

Stillingfleet has generally been recognized as correct in his claim that a philosophical defense of Anglican Protestantism would require a concept of substance to support the mystery of the Trinitarian doctrine of three distinct attributes or natures present in one God. Locke's agnosticism regarding substance in the *Essay* did mean, as Locke paraphrased Stillingfleet's charge against him, that he had "almost discarded substance out of the reasonable part of the world." But Locke did not want to admit this. He insisted that he had written merely that our ideas of substance were "obscure and relative," which did not mean that substance itself was obscure and relative.[18] But if all of our knowledge comes to us from our ideas and our reflection upon them, it follows that we have no knowledge of substance itself. And if we have no knowledge of substance, then substance cannot be part of an empirical world view.

Locke used a similar hedge to answer Stillingfleet's main objection to his location of forensic personal identity in a "same consciousness" come the Resurrection. Stillingfleet insisted that according to Locke, at the Resurrection the same person would not need to have the same body that he occupied while alive. The full force of Locke's theory of personal identity discussed in chapter 5 remains opaque until Stillingfleet's theological criticism is considered. Stillingfleet argued that Christian doctrine requires a physical resurrection of the dead, even though he was well aware that this did not simplistically mean that the literal contents of buried coffins would be reanimated. Stillingfleet meant that a concept of substance, perhaps allowing for essential characteristics of an individual's body, or the "form" of his or her body, was necessary to make common sense of the Resurrection doctrine.[19] Locke, who had reduced the immortal personal *agent* to conscious awareness and awareness of its continuity through memory, reduced to absurdity Stillingfleet's objections to his theory of personal identity by speculating on what the resurrection of the body according to scripture could not mean, and by making it seem as though this was what Stillingfleet had in mind:

> But setting aside the substance of the soul, another thing that will make any one doubt whether this your interpretation of our Savior's words be necessarily to be received as their true sense, is That it will not be very easily reconciled to your saying, you do not mean by the same body the same individual particles which were united at the point of death. And yet, by this interpretation of our Savior's words, you can mean no other particles but such as were united at the point of death; because you mean no other substance but what comes out of the grave; and no substance, no particles come out, you say, but what were in the grave; and I think your lordship will not say, that the particles that were separate from the body by perspiration before the point of death were laid up in the grave.[20]

Locke is here implying that a notion of physical substance is too broad to support the resurrection claims of scripture because it would include bodily particles, such as evaporated perspiration, that could not be present in the grave. His philosophical point would be that substance is a contradictory notion because it commits us to including ideas of parts of an object in our meaning of that object, which parts are not intended in our meaning. At the beginning of the reply, Locke made it clear to Stillingfleet that he believed in the Christian faith, based on revelation.[21] But, as commentators have pointed out, revelation was acceptable to Locke only if it introduced no new simple ideas that had not been present before the revelation.[22]

It is difficult to avoid the impression that Locke not only triumphed over Stillingfleet, who died before their controversy was resolved, but that he enjoyed outwitting the bishop and twisting the knife in front of third parties. For instance, in his "Second Reply," he claimed that Stillingfleet misused language. He then wrote a memorandum to John Freke, the "bachelor" and host to the "College," a political club through which Locke made his ideas known to members of Parliament:

> But a professor to teach or maintain truth should . . . speak plain and clear and be afraid of a fallacy or equivocation, however prettily it might look and be fit to cheat the reader, who on his side should be an author who pretends instruction abominate all such arts and him that uses them as much as he would a common cheat who endeavours to put off brass money for standard silver.[23]

Since Locke had not been candid with Stillingfleet, the wit in the above passage may have included a warning about himself that he could afford to make to his audience at that time. And in light of Locke's important contribution to financial policy—he had recently been instrumental against depreciating the coinage—the simile of currency probably evoked memories of his own worth.[24]

The Pragmatics of Toleration

There are various reasons for tolerating diverse religious beliefs and activities: people have an absolute right to religious freedom, or are entitled to it as part of their other absolute rights; religious agreement is too difficult to impose; peace and happiness at large require toleration; toleration is constitutionally written into a foregone separation of church and state, as in the American tradition. The American tradition was the result of arguments for toleration that were formulated before there was a clear separation of the powers of governmental and religious institutions. Throughout the seventeenth century in England, different positions on religious toleration were imbedded in religious perspectives and persuasively advanced in appeals to the self-interest of those who were opposed to toleration. Although Locke's foundational views on toleration are often understood as part of a continuum that began with John Milton's views, it is important to realize that Milton was mainly concerned with freedom of the press, whereas Locke's views entailed the broader doctrine of a separation between church and state. Milton's arguments against censorship were based on the value of learning and free inquiry, whereas Locke's views were based on his commitment to achieving a prosperous, stable, and peaceful society.

Milton addressed the *Areopagitica* to Parliament in 1644, after his unlicensed pamphlet in favor of divorce had been attacked and censured as a "wicked book."[25] He argued against censorship on historical, religious, and moral grounds: except for the Catholic Counter-Reformation, censorship has always been a historical anomaly in cultures that value learning; God entrusted adult men with rational choice; virtue has to be tested by exposure to learning. Milton also argued that censors themselves are likely to err, that people can be led astray without books, and that the regulation of reading could lead to the regulation of all pleasurable recreations, including Christian meetings.[26] He insisted that reason, like virtue, requires constant choice, and that it is injustly humiliating to scholars and writers to require that their books be licensed. He claimed that uncensored discussion is the best way to encourage the development of new ideas and it is more likely than imposed religious pieties to support the search for truth.[27]

Milton was addressing the 1637 decree of the Star Chamber. Even though the Court of the Star Chamber had been abolished in 1637, Parliament had taken over its office. Milton's efforts were not successful. However, fifty years later, Locke, through his influence with Edward Clarke and John Freke, who had been appointed members of a House of Commons committee in 1694–95, did succeed in getting the Act for the Regulation of Printing repealed. Locke's conclusive arguments were that the monopoly exercised by the Company of Stationers, under the Act, was a restraint of trade opposed to the public interest.[28]

Locke's arguments in the *Letter Concerning Toleration* and its defenses encompassed civic as well as religious liberties in ways that reflected his concerns about the political turmoil of his times. Locke was chiefly concerned with economic prosperity, and perhaps with freedom of scientific inquiry as well. James Tully points out that from before the Civil War until 1689, the issue of religious toleration swirled around the efforts of Puritan, mild Anglican, and radical dissenters to gain political power, variously against the monarchy, Parliament and the established Church of England. Freedom of religious action was a result of political power. If an individual had full civil liberties he would automatically have religious freedom because the dominant religious group effectively ruled the country and those who ruled the country imposed their religion on those who were ruled.[29] Thus, religious toleration and political toleration were the same issue. From the standpoint of religious dissenters, arguments for religious toleration were arguments for their political rights or for an increase in their political power.

Locke's first writings about toleration were pragmatic arguments about how best to keep the peace. In 1660 and 1661, in two tracts, he reasoned that a monarch (i.e. Charles II) had the right to refuse toleration to dissenters for the public good.[30] But then, in 1665, during a diplomatic visit to Cleves, he saw Dutch Calvinists, Lutherans, and Catholics coexisting in harmony.[31] When he returned to England, he became allied with Anthony Ashly Cooper, Lord Shaftesbury. Thereafter, Locke's writings on toleration began to incorporate the importance of subjective religious choice. The fining, imprisonment, and exile of 15,000 Quakers and the death of 450 of them during the enforcement of the Clarendon Code in the 1660s made it clear that it was impossible to expect dissenters to exercise spiritual choice only, while conforming outwardly to an imposed religion that went against their beliefs. Religious conscience among dissenters required that they openly oppose the law and suffer the punishment.[32] Lord Shaftesbury was himself a Presbyterian and it is understood by even traditional historians that his support of toleration had little to do with his personal spiritual commitment but was rather connected with his vision of increased prosperity through colonial expansion and international trade.[33] In order to support these activities, mercantilism required unity on a national level, and internal religious fighting worked against this.

Tully notes that in the *Letter Concerning Toleration,* written in 1685, Locke was referring to Charles II's reprisals after the defeat of the Monmouth Rebellion of 1685, a persecution that entailed a mass hanging of 150 dissenters and the imprisonment and execution of many others. After Lord Shaftesbury's death in 1682, Locke became involved in Algernon Sydney's plot to assassinate Charles in 1683. When Sydney was hanged, Locke fled to Holland, where he helped organize the Monmouth Rebellion.[34]

Locke wrote the *Letter* in Latin, but after its 1689 publication it was quickly translated into English by William Popple, who commended it to his countrymen for immediate practical use.[35] Although Charles had by then been succeeded by James II and William of Orange, the issue of dissent and toleration was still relevant. The Toleration Act of 1689 denied freedom of worship to those who denied the doctrine of the Trinity; Trinitarian Protestant dissenters could get a license to meet if they took an oath of allegiance, but they could not hold public office.[36]

Locke begins the *Letter* with a forceful claim that toleration is "the chief characteristical mark of the true church" because the business of religion has to do with regulating "mens lives according to the rules of virtue and piety." Controlling immorality in others through the "extirpation of sects" is not a religious task beause love and not force is the only religious means of persuasion. Salvation is an individual obligation and therefore not the duty or right of civic rulers to ensure or enforce. Rulers can use persuasion if they have a better idea than their subjects about how to attain salvation, but they are then on the same level as all other citizens and preachers.[37]

Locke shrewdly observes that laws have no force without penalties and then he argues that penalties "do not convince the mind." He defines a church as a free and voluntary society that thereby has the right to make its own rules. Such rules are enforced by exhortation and, in the last resort, expulsion, but force is inappropriate because all force is in the domain of the magistrate or ruler, that is, the civil government. Whatever is legally allowed or forbidden in nonreligious matters should be allowed or forbidden in worship. For example, if people can eat, wear, and say what they want in their secular lives, then they should have the same freedom in worship; if it is legal to kill a calf for ordinary purposes, then it should be legal to kill a calf for God.[38]

Locke's justification for the institutional division of church and state is that it is the only way to secure peace:

> No body therefore, in fine, neither single Persons, nor Churches, nay, nor even Commonwealths, have any just Title to invade the Civil Rights and Worldly Goods of each other, upon pretence of Religion. Those that are of another Opinion, would do well to consider with themselves how pernicious a Seed of Discord and War, how powerful a provocation to endless Hatreds, Rapines, and Slaughters, they thereby furnish unto Mankind. No peace and Security, no not so much as Common Friendship, can ever be established or preserved amongst Men, so long as this Opinion prevails, That *Dominion is founded in Grace,* and that Religion is to be propagated by force of Arms.[39]

According to Locke's writings in the *Second Treatise on Government,* "dominion" cannot be founded in "grace" because the basis and justification for government is the preservation of the life, liberty, and estate of its citizens. The upshot of his arguments for toleration in the *Letter* is that religion is not the concern of magistrates, who have no right to concern themselves with the salvation of souls.

It is doubtful that any astute observer of seventeenth-century English politics genuinely believed that religious persecution was undertaken with the motive of saving souls. However, Locke shored up his argument that rulers ought not to impose religion with an empirical claim that it was impossible to do so. He argued in effect that religious belief is by definition a private, subjective matter:

> Whatsoever may be doubtful in Religion, yet this at least is certain, that no Religion, which I believe not to be true, can be either true, or profitable unto me. In vain therefore do Princes compel their Subjects to come into their Church-communion, under pretence of saving their Souls. If they believe, they will come of their own accord; if they believe not, their coming will nothing avail them. How great soever, in fine, may be the pretence of Good-will, and Charity, and concern for the salvation of mens Souls, men cannot be forced to be saved whether they will or no. And therefore, when all is done, they must be left to their own Consciences.[40]

This idea of completely private and nonpolitical conscience and responsibility complements Locke's theory of personal identity, as discussed in the previous two chapters. The religious individual is a sealed-off self, open only to God "on that Great Day," and he owns not only the past memories of himself as an agent but his own privacy. It is difficult to determine how much of this picture was for Locke fundamental Protestant religious belief, and how much was rhetoric to secure social peace and security through religious tolerance. He argued elsewhere in the *Letter* that children do not inherit their religion from their parents, though it must have been the case in his time, as in ours, that most people do have the same religion as their parents.[41] He must have realized that not all religious belief is freely acquired, that people can be frightened into accepting creeds or have superstitions imposed on them. His model for the believer seems to be a rational adult, probably male, and someone not given to self-deception, though it is not clear that Locke had a concept of self-deception. Locke assumes that those whose beliefs a ruler might try to control are individuals who are otherwise worthy of respect. His argument does not apply to those who antecedently are not candidates for full civil liberties because there would be no practical reason for an unjust ruler to grant tolerance or recognize freedom of conscience in such individuals.

However, Locke's final argument for toleration in the *Letter* is that oppression of any kind occasions sedition because it gives people common ground on which to unite against their oppressors. This is a difficult argument to evaluate if oppression seems to be effective at any given time. Tully here interprets Locke as using revolution as a threat against injustice.[42] But Locke can also be read as making the simple claim that peace begets peace and all acts of force only lead to forceful reactions.

Locke's arguments in the *Letter Concerning Toleration* were used in later centuries to support doctrines of individual rights and the separation between church and state. But at the time, his Anglican critics continued to insist on their right to use force against dissenters—probably because they already had the balance of power and could effectively impose their doctrines on dissenters.[43]

Returning to my claim at the outset of this chapter that religion in seventeenth-century England could not have furnished institutional support for science, it should now be clear that religion was at the time the medium for the most extreme forms of political conflict. Except for Milton's writings on censorship, arguments about religious toleration had little to do with free inquiry in general. Besides, even the probablist epistemology of liberal Protestants was too metaphysical (in its reliance on substance) to be compatible with philosophical empiricism. As an activity, science would have to find its own wherewithal, and that, ironically, made it a refuge from religious contention. As Shapiro points out, many Latitudinarians saw their participation in the Royal Society as a respite from politics, as well as religion.[44]

Shapiro assimilates the free inquiry of the Royal Society to Latitudinarian toleration in general, and many other writers have drawn broad historical connections between English Protestantism and English empiricism.[45] However, it is important to remember that the doctrine of the Trinity, insofar as it required a concept of substance, made impossible metaphysical demands of empiricism, as the Locke–Stillingfleet controversy brought out. Whether Locke's philosophical skepticism about substance was motivated by his rejection of the doctrine of the Trinity (on either religious or political grounds) is an interesting question. But the question of religious toleration did not extend to natural philosophy, except indirectly in the issue of book licensing. It was not necessary for Milton or Locke even to mention science in their writings on censorship or toleration, and they did not. This freedom to pursue natural philosophy regardless of other disputes was not only important for the development of science, but also emphasizes the separation of scientific thought and practice from both religion and politics in England.

I have already suggested that science was an enjoyable recreation for the bachelors of science. Through its earliest associations with "the betterment of mankind" in Bacon's optimism about technology, science in England seemed to need no external justification. It also became self-ruling, as I shall explore in the next two chapters about the Royal Society and Isaac

Newton. The seventeenth-century separation of science from other strife-torn spheres of thought came to provide the model for intellectual and academic freedom generally. And under that model, the seventeenth-century bachelors of science liberated themselves. As the natural philosophers (i.e., scientists) and virtuosi of the Royal Society who invented modern science, they invented their own liberated roles in it, and thereby invented themselves. Had they not been, partly deliberately and partly inadvertently, so exclusive, this self-invention could have liberated many diverse "others" through free, interesting work.

Eight

The Royal Society

Nullius in Verba

—MOTTO OF THE ROYAL SOCIETY

One of the strongest, if still unwritten, rules of scientific life is the prohibition of appeals to heads of state or to the populace at large in matters scientific. Recognition of the existence of a uniquely competent professional group and acceptance of its role as the exclusive arbiter of professional achievement has further implications. The group's members, as individuals and by virtue of their shared training and experience must be seen as the sole possessors of the rules of the game or of some equivalent basis for unequivocal judgments.

—THOMAS S. KUHN, *The Structure of Scientific Revolutions*

The Royal Society of London for the Improvement of Natural Knowledge provided respite from the political disputes that were closely tied to religious differences in England during the middle of the seventeenth century. It was the medium through which some types of philosophical inquiry changed into empirical sciences—as late as 1887, the official publication of the Royal Society was called *Philosophical Transactions.* In this chapter and the next, I discuss the intersection between the institutional character of the Royal Society and the character of its individual participants in terms of the resulting identities of scientific knowers. Building on Thomas Kuhn's distinction between the Baconian sciences and the Newtonian system, I examine two main types of scientific identity: the first, humbly tentative, inductive, and fallible; the second, robustly confident, deductive, and unerring. The first type, which was, oddly enough, Baconian, will be the primary focus of this chapter. We will see in the next chapter that the

Newtonian type of identity was sufficiently grand as to vault over the human intellectual virtues into divinity.

The first section of this chapter is a brief history of the origins of the Royal Society. An analysis of the Hermetic antecedents of its Baconian inductivism follows. Boyle's elimination of Hermeticism from empiricism is then noted. Finally, a model of the resulting empirical knower and nonknower, which is relevant to liberatory concerns today, is entertained.

Historical Origins

It is frequently reported that the immediate precursor of the Royal Society was the "invisible college" that met in Oxford at Wadham College during the 1650s. However, there was half a century of institutional and ideological background to the invisible college, which few historians of ideas take the time to delineate. The following account is drawn from Dorothy Stimson's careful description of the antecedents to the Royal Society in *Scientists and Amateurs.*[1]

Francis Bacon's *New Atlantis* was first published in 1626 and went through ten editions by 1670. The centerpiece of this utopia was the House of Solomon, a research institute with laboratories for experiment and observation in the natural sciences that were to include studies of heat, light, cold, medicine, minerals, weather, crafts, astronomy, animals, and agriculture. Among the staff of thirty-six fellows and their assistants would be "merchants of light" who would spend years abroad collecting new observations, and scholars who would critically read all past written works in search of useful experiments. Three "Interpretors of Nature" would evaluate the work of all the others in order to abstract axioms and principles of natural knowledge. Bacon's scientific readers have honored this vision of the House of Solomon, down to portraying him on the frontispiece of virtually all published official histories of the Royal Society.

Comenius, a Czech, envisioned an educational system, similar to Bacon's, that was epitomized by a "pansophic" university of all knowledge for the systematic study of nature. Comenius came to London in 1641 in hope of raising financial backing from English nobles and Parliament, but left disappointed. His sponsor was Samuel Hartlib, a merchant born in Elbing, who was a well-regarded "intelligencer" of science during the years of Charles I. Hartlib's *Description of the Fameous Kingdom of Macaria* provided for a center of practical learning. William Petty, Hartlib's friend, who was physician-general to Cromwell's army in Ireland and a founder of modern economics, also envisioned a school for practical education that would include histories of the various trades for making a living. John Evelyn, diarist, author of *Sylva* (on forest trees), and commissioner of public services to veterans and prisoners of the Dutch War, proposed a practical learning center on an estate outside of London to his friend, Robert Boyle. Abraham Cowley's 1661 *Proposition for the Advancement of Experimental Philosophy* describes a college of twenty philosophers and sixteen assistants for the practice and growth of useful knowledge, as a correction to the theoretical curricula of Oxford and Cambridge.

These ideas of Bacon, Comenius, Hartlib, Petty, Evelyn, and Cowley already had a real foundation in Gresham College. Thomas Gresham, a merchant and Elizabeth I's financial agent, had bequeathed his and his wife's estates to the City of London and the Mercers Guild, to provide citizens with lecturers and a place to meet. Beginning in 1598, the seven unmarried professors of Gresham College lectured on law, physic, rhetoric, divinity, music, geometry, and astronomy, in English as well as Latin (the sole language of Oxford and Cambridge professors). Their audience consisted of scholars, nobles, and business and professional men. The Gresham professors, who received comfortable living quarters and £50 a year, included Henry Briggs, developer of logarithm tables; William Oughtred, inventor of slide rules; Seth Ward, later bishop of Exeter and Salisbury; and John Wallis, who chronicled the foundations of the "invisible college."

During that time, many nobles and prosperous Puritans were beginning to collect ancient relics and other curiosities, and there was popular interest in "mechanical marvels." John Wilkins, in *Mathematical Magic,* described the "admirable contrivances" of natural things, such as the eye of a fly he had seen through a compound microscope. Wilkins was a strong proponent of the Baconian thesis that nature is a second Bible and natural philosophy a productive form of worship.

During the early decades of Gresham College, many European Protestants who were also interested in the new sciences took refuge in London, and English royalists, who were exiled on the continent during the Civil War period, traveled to places of learning there. Thus, the association of interest in the new sciences with peace and material comfort had an international dimension.[2] The Protestant virtues of industriousness and the accumulation of wealth were to be fulfilled by technological progress from the new science. This connection has been well studied by later modern historians of ideas.[3] But the emphasis on the potential lucrativeness of practical knowledge in the seventeenth century sometimes obscures the more uninterested pursuit of this knowledge, even by merchants.

Theodore Haak, a Calvinist originally from Worms who worked as a language translator, suggested the formation of the invisible college. A group of younger scientific amateurs that included Robert Boyle, who was then eighteen, were meeting weekly in London by 1645. Boyle referred to their discussions of scientific news over lunch as the "invisible college" because there were no buildings or professors. According to Wallis, recollecting in 1678, discussion of religion and politics was against the rules and subjects were confined to "Philosophical inquiries and such as related thereunto; as Physick, Anatomy, Geometry, Astronomy, Navigation, Statics, Mechanicks and Natural Experiments."[4] They met in taverns, at members' houses, and at Gresham College, which was also to house the Royal Society. They discussed the Copernican theory, William Harvey's evidence for the closed circulation of the blood, barometric experiments with mercury ("quick silver"), and William Gilbert's study of magnetism.[5]

After Charles I's execution, Wallis, Jonathan Goddard, and John Wilkins replaced royalists in academic posts at Oxford and the invisible college partly reassembled there, as some of the 1645 membership organized the Philosophical Society of Oxford in 1651. They kept records and attempted to collect dues. Wilkins was warden of Wadham College at that time, and Boyle moved to Oxford during the 1650s, as did Seth Ward, who became Savilian Professor of Astronomy. Thomas Sprat, the future historian of the Royal Society, and Christopher Wren, astronomer and architect, also attended. Evelyn described Wilkins's generous hospitality to the group as well as his collection, which included transparent apiaries and a hollow statue that "spoke" through a concealed pipe. Wilkins married Oliver Cromwell's younger sister and was made master of Trinity College at Cambridge, but he was sufficiently skillful or apolitical to regain royal support after the Restoration. Indeed, Wilkins united religiously tolerant Latitudinarians, as well as the new scientists.[6]

Following a lecture on astronomy by Christopher Wren in 1660 at Gresham College in London, those in attendance decided to found a college "for providing Physio-Mathematical learning." Within a week, Charles II approved. Officers were agreed upon, notes would be taken, and a charter book was set up. The original membership of 115 included those from the first invisible college, Wilkin's Wadham group, Gresham College professors, doctors, divines, lawyers, literary men, civil servants, and ten noblemen, as well as the king. Only a third were scientists, and Lord

Brouncker, the leading mathematician, was the first president. Evelyn named it the Royal Society. The charter passed the Great Seal on July 15, 1662, and Charles presented the Society with a silver mace. The king was not only intrigued by the new science—he had maintained a laboratory himself during his years of exile—but enjoyed making fun of the more preposterous experiments undertaken by its practitioners.[7]

The continuity in membership from the invisible college and the support and good will of the king and Gresham College suggest that the Royal Society began as a relaxed, convivial association. The decades of informal meetings and association that preceded the Society justify viewing it as the culmination of what in the twentieth century would be called a "movement."

The Baconian Sciences

As a collective enterprise, the Baconian sciences were miscellaneous and pluralistic. Thomas Kuhn has emphasized their incommensurability with the "mature" sciences that eventually supplanted them, because the "winning" modern scientific paradigms involve radically different semantic perspectives that went beyond mere falsification of their predecessors.[8] Even though Kuhn's sympathetic interpretation of the relativity of the truth claims of ancient, medieval, and early modern scientific theories does not fully dislodge the sense of their "immaturity" in the history of science, the Baconian sciences support a positive reading of the epistemic identities of their practitioners. On the way toward that positive reading, the nature of the Baconian sciences and their transformation through Boyle's corpuscularian theory are discussed in this section and the next.

As Kuhn explains, the classical sciences from antiquity were astronomy, statics, and optics, all of which were associated with mathematics and harmonics. There was a tradition of unified study across these five subjects because advances in one led to advances in the others through their common deductive structures, which were all closely related to, when not coincident with, geometry. The factual base of these sciences was ordinary observation. During the sixteenth century, the study of local motion, as distinct from Aristotle's philosophical concept of qualitative change, was added to the mathematical sciences. In the seventeenth century, all of the mathematical sciences were fundamentally revised through the addition of analytic geometry and calculus; the substitution of the heliocentric theory for the geocentric theory; the development of quantitative laws of motion; new theories of vision, refraction, and colors; and the extension of statics to pneumatics.[9]

Kuhn argues that, contrary to "seventeenth century rhetoric," both Aristotle and the medievals recognized the importance of observation. However, it was not always clear which reported experiments in those traditions were thought-experiments and which had actually been performed. Those that were performed tended to demonstrate foregone conclusions or answer questions posed by accepted theories. By contrast, the Baconian scientists conducted experiments to determine what would happen under new conditions. Practitioners of the Baconian sciences also created conditions for observation that could not have occurred naturally, for example, putting small animals in barometers. A third major departure from classical methodology was the development and use of instruments such as telescopes, microscopes, thermometers, barometers, air pumps, and electric charge detectors. Finally, a premium was placed on the actual performance of experiments before witnesses.[10]

Kuhn agrees with Alexandre Koyré and Hubert Butterfield that these empirical innovations did not so much lead to unanticipated discoveries as reflect and promote "new ways of looking at old phenomena." That is, the methods of Baconian empiricism may have contributed very little to the subject matter of the classical sciences. But the Baconian empirical innovations gave rise to new fields such as magnetism, electricity, heat, and chemistry, which developed as a result of the use of new instruments and studies of existing crafts. However, except for England, where the classical sciences declined in the second half of the seventeenth century and Newton combined both classical subject matter and discoveries in the Baconian sciences, practitioners of the new fields were not recognized institutionally by classical scientists until the nineteenth century. Except for the chemists who were employed in medicine and industry, these new scientists were mainly amateurs. For example, Guillaume Amontons, who contributed to the theory and design of the thermometer and hygrometer, never rose above *élève* in the French Academy, and Pierre Polinère, who introduced *physique experimentale* to France, was never part of the Academy.[11]

During the first third of the seventeenth century, the Baconian sciences were strongly influenced by occult Hermetic concepts of natural sympathies and antipathies. This magical and at times even animistic outlook was conducive to the nonmechanical and nonmathematical crafts of medicine, dyeing, glass-making, navigation, and chemistry, which provided data for

Baconian science. The crafts were associated with the figure of a *magus artificer,* who magically manipulated nature for human benefit. Several writers, including Kuhn, have related the utilitarian values of Puritan amateur scientists to Hermeticism through a series of external connections:[12] The Puritan virtue of hard work was associated with the goals of the crafts to make money and make human life more amenable. Masters of the crafts were also associated with beneficent magi. Practitioners of the Baconian sciences sometimes had Hermetic beliefs, and many of them were also Puritans.

An internal intellectual connection between mysticism and Baconianism is drawn by P. M. Rattansi, who traces the Hermetic strains in English mid-seventeenth-century science to Giordano Bruno, a late sixteenth-century Dominican heretic who developed mystical Copernicanism. Tommaso Campanella, following Bruno, wrote in *City of the Sun* of a utopia where scientific practice and astral magic would be encouraged for the good of the community. Comenius's ideas were in this school of thought, and, as noted in the previous section, Hartlib was instrumental in sponsoring Comenius in England.

Robert Boyle and William Petty were both impressed by Hartlib's humanitarian visions. The main goal of alchemy by then had changed from the transmutation of base metals into gold to the prolongation of human life through new discoveries in medicine. Boyle doctored himself and collected medical prescriptions. He was a chemist (as was Hartlib's son), and after he came in contact with Gassendi's atomism and Cartesian mechanism, he developed his corpuscular philosophy of science. Rattansi therefore interprets Boyle's corpuscularianism as an attempt to reconcile Hermeticism with empiricism.[13]

Boyle's Revision of the Baconian Sciences

At first, Boyle was critical of atomism on religious grounds. He began to develop the corpuscularian and mechanical theories when he joined Wilkins's group at Wadham in the 1650s. By that time, Walter Charleton had taken up Gassendi's ideas in preference to J. B. van Helmont's Hermeticism. Henry Moore attempted to combine neo-Platonism with Cartesianism, while Seth Ward was expressing concern about the enthusiasm of radical political groups for a universal Hermetic curriculum. Boyle was worried about religious factionalism as the number of Protestant sects multiplied. According to many of these sects, the term *experiment* included mystical illumination. The Cambridge Platonists, who had evolved into Latitudinarians, generally saw the varied forms of religious enthusiasm as a threat to social stability and to mild Anglican toleration. Rattansi argues that Boyle did not relinquish Hermeticism, at least not in his early exposition of the mechanical philosophy. For example, Boyle claimed that the growth of plants and of metals and gems in the ground could be explained only by "seminal forms."[14]

However, against Rattansi's somewhat circumstantial case for Boyle's Hermeticism, Kuhn notes that the nonmechanical sciences were generally less amenable to the corpuscular theory because they were nonquantifiable.[15] Presumably, what later became geology and botany would have had that deficiency in the mid-seventeenth century. Since chemistry was the Hermetic inquiry par excellence, however, it is puzzling that Boyle, a chemist, should have been the main English proponent of the corpuscularian theory because corpuscularianism could not be applied to chemistry at that time.

On a textual basis, Boyle's positive requirements for the accumulation of scientific knowledge support an anti-Hermeticist interpretation. In *The Skeptical Cymist,* Boyle's main objection to Parcelsus's and van Helmont's theories of elements was that they were not supported by reported experiments with fire.[16] In his 1661–63 correspondence with Spinoza, which was mediated by Henry Oldenburg when he was secretary of the Royal Society and Boyle was president, it is evident that Boyle considered observation and experiment to be the basis of science. As A. Rupert and Marie Boas Hall put it:

> Is a scientific proposition sufficiently demonstrated when it is shown to be a logical consequence of a set of intuitively certain axioms? Descartes would have answered affirmatively, Boyle negatively. Is a scientific proposition sufficiently demonstrated only by showing empirically that it holds? Descartes would have answered negatively, Boyle affirmatively.[17]

In a letter to Oldenburg, in response to Boyle's *Certain Physiological Essays,* Spinoza wrote that experiments could never prove the particulate (atomic) structure of matter because only reason and calculation could "explain Nature as it is in itself, and not as it is presented to the human senses."[18]

Boyle's position was compatible with Spinoza's in the same way that the Gassendi-Mersenne *via media* skeptical epistemology was compatible with Catholic doctrine (as discussed in chapter 3). Boyle would not have tried to use empirical findings to explain nature as it was in itself, but, unlike Spinoza and Descartes, neither did he accept deductive conclusions about nature as a form of scientific knowledge. Like Gassendi and Mersenne, Boyle (skeptically) relinquished the possibility of knowing ultimate truths about nature. He sought empirical—that is, experimental—confirmation of the corpuscular theory because he did not think that having a concept of something was proof of the reality of that thing. But Boyle's insistence on experimental confirmation did not entail that the confirmation constituted real knowledge about hidden things. On the contrary, for Boyle it entailed that there is no knowledge, scientific or otherwise, about hidden things. That position puzzled Spinoza because of the priority he gave to reason.[19] Spinoza, in line with Descartes, would have assumed that reason could yield knowledge about hidden things.

Boyle did not give priority to reason as a way of gaining knowledge about the world. He even took Pascal to task for relying on thought experiments and presenting diagrams of scientific instruments that could not have been constructed.[20] Boyle insisted on actual observation of experiments. Descriptions of such observations could support hypothesis and theory, but he would "have such kind of superstructures looked upon only as temporary ones; which though they may be preferred before any others, as being the least imperfect, or, if you please, the best in their kind that we yet have, yet are they not entirely to be acquiesced in, as absolutely perfect or uncapable of improving alterations."[21]

Even Boyle's most direct arguments for the preferred corpuscular theory were ultimately heuristic. In *The Excellency and Grounds of the Corpuscular or Mechanical Philosophy,* he argued that the assumption that natural phenomena can be explained by matter and motion is the most simple or general theory, such that "These principles, matter, motions (to which rest is related), bigness, shape, posture, order, texture, being so simple, clear and comprehensive, are applicable to all the real phaenomena of nature, which seem not explicable by any other not consistent with ours."[22]

The Humble Knower

Traditionally, philosophers and historians of ideas have viewed Boyle's type of inductive uncertainty as a necessary price for empirical knowledge. As discussed in chapter 2, Richard Popkin locates this uncertainty in a wider intellectual context of seventeenth-century theological skepticism.

Barbara Shapiro, on the other hand, identifies the fellows of the Royal Society in the 1660s as a "new intellectul type" due to their search for and acceptance of what was "probably true" or no more than "morally certain."[23] Both of these accounts neglect the intellectual identity of a knower who is prepared to accept the limited and temporary nature of what can be known. Such a knower must constantly have an idea of what is unknown in order to carry on effectively in situations of knowledge. He or she must know what he or she does not know and accept ignorance rather than deny it or structure epistemic claims to assure that what little is known is "good enough." This kind of humility seems to have been deliberately cultivated as an intellectual virtue during the pre-Newtonian period of the Royal Society. In his 1667 *History of the Royal Society,* Thomas Sprat outlines the intellectual virtues of a *virtuoso:*

> *The Natural Philosopher is to begin, where the Moral ends.* It is requisite, that he who goes about such an undertaking, should first know himself, should be well-practis'd in all the modest, humble, friendly Vertues; Should be willing to be taught, and to give way to the Judgement of others. And I dare boldly say, that a plain, industrious Man, so prepar'd, is more likely to make a good Philosopher than all the high, earnest, insulting Wits, who can neither bear partnership, nor opposition. . . . For certainly, such men, whose minds are so soft, so yielding, so complying, so large, are in a far better way, than the Bold and haughty Assertors: they will pass by nothing, by which they may learn: they will be always ready to receive, and communicate Observations: they will not contemn the Fruits of others diligence: they will rejoyce, to see mankind benefited, whether it be by themselves, or others.[24]

These ideals of open-mindedness, cooperation, and goodwill toward colleagues for the wider benefit of society presuppose an admission of ignorance on the part of a knower. If nonassertive language and a refusal to accept the word of authority are also in play, then a sense of shared, individual ignorance emerges that requires tolerance of different knowledge claims. There is something refreshing about such intellectual humility as a virtue. If virtues are components of identity, and if a capacity for knowledge or knowing is a virtue, then the virtue based on intellectual fallibility is a fissure—a contradiction—in one's identity as a knower. The contradictory combination of being a knower and someone who doesn't know allows for the possibility of learning—one may come to know anything at all as a result of that contradiction. And the identity of the knower-non-knower is even shakier if the objects of knowledge are no more than probably true.

Of course, even genuine ignorance, like pyrrhonic skepticism, can be part of a "machine of war" against opinions one wants to demolish to the benefit of one's own "hypotheses." The founders of the Royal Society were not unworldly: Royal patronage was sought as protection from anticipated religious opposition; much of Sprat's history is an apology of the new science addressed to possible charges of impiety; the two-thirds membership that was honorary to the nobility or made up of amateurs was accepted by scientists to increase funding.[25] But if the humble type had been put forth as as a humbug for ulterior reasons, one would expect a background agenda to move scientific truths already accepted as certain to intellectul dominance, and there is no evidence of such plans in the early days of the Society.

Robert Hooke, after assisting Boyle with his chemical research for eight years, began his forty-year career as Curator of the Royal Society. Hooke wrote out the aims of the Royal Society when the Second Charter was granted in 1663, and he claimed that its business and design was to:

> examine all systems, theories, principles, hypotheses, elements, histories and experiments of things naturall, mathematicall and mechanicall, invented, recorded, or practised by any considerable authors ancient or modern. In order to the compiling of a complete system of solid philosophy for explicating all phenomena produced by nature or art, and recording a rationall account of the causes of all things. In the meantime this Society will not own any hypothesis, system or doctrine of the principles of natural philosophy, proposed or mentioned by any philosopher ancient or modern, nor the explication of any phenomena whose recourse must be had to originall causes (as not being explicable by heat, cold, weight, figure and the like effects produced thereby); nor dogmatically define nor fix axioms of scientificall things, but will question and canvass all opinions adopting nor adhering to none, till by mature debate and clear arguments, chiefly such as are deduced from legitimate experiments, the truth of such experiments be demonstrated invincibly.[26]

There is no expectation, in Hooke's prospectus, of a paradigm in the Kuhnian sense. It might be countered that within the specialized sciences there were paradigms, such as theories of Heliocentrism, corpuscular mechanism, and the closed circulation of the blood. But insofar as all fellows and foreign members of the Society were the audience for all of its work, and dissent was expected and accepted, there could be no general paradigm. In that respect, it is not a misnomer to refer to the virtuosi as philosophers—even though many contemporary philosophers are as yet unwilling to acknowledge or support a similar aparadigmatic pluralism in their own field, today.[27]

At this point something needs to be said about the "homosociality" of the early fellows of the Royal Society. The term homosociality comes from recent literary critical scholarship, by Eve Sedgwick and others, according to which close affective ties among men are the real dynamic behind relationships between men and women.[28] Given the collegiality among the virtuosi, a case could be made that their relationships with one another gave rise to the motivations for their scientific inquiries. This might suggest that the real subject matter of the early Royal Society was not observation, experiment, and the collection of natural histories but the winning of affect and power among men. The kind of biographical research necessary to support or refute such a claim could be found in that new web of personal and power relationships which must have been coincident with the beginnings of the modern scientific project. The relevant studies are beyond the present scope but the questions for inquiry are fairly obvious: What were the major friendships, rivalries, and jealousies? How did social interaction intersect with scientific projects? Who were the elite and how was their influence effected? Answers to such questions need not entail dismissal of the work of the virtuosi or their scientific claims. Rather, the homosocial explanation could complement the intellectual history account.

The issue of the exclusion of women from the fellowship and audience of the Royal Society, with the sole exception of the anecdotal instance of Margaret Cavendish, which I will get to soon, now needs to be addressed. I argued in chapter 1 that the misogynism of seventeenth-century empiricists was de facto—women were excluded from scientific enterprise as they were excluded from all distinguished employment. It seems plausible that Bacon's masculinization of science was a rhetorical tool to recruit men by the use of popular sexual imagery. There was not yet any widespread gender identification of women as incapable of intellectual work. Still, the exclusion of women has been widely read by feminist writers as inherent to the nature of the seventeenth century empirical movement, which they see as a men's movement. (For example, Linda Jean Shepherd, in *Lifting the Veil,* quotes reiterations of Bacon's masculinizing rhetoric by Henry Oldenburg and Joseph Glanvill.)[29] Indeed, the masculine character of the new sciences was emphasized throughout the writings of the first members of the Royal Society, including, of course, Thomas Sprat. Nonetheless, the earlier point holds. At the beginning of the Royal Society, especially in the type of the humble and self-confessedly ignorant scientists, it is evident that the primary concern was not with the traits of women, who were excluded, but with constructing the identities of those who were included.

This means that the original de facto exclusive presence of men in science *scientized masculinity* as much as it may have *masculinized science.*

The scientization of masculinity created a new, privileged identity for some early modern men. This new identity did not extend to all early modern men: Great landowners and nobles did not need it (though they might take it up for recreation), and it was beyond the reach of men without leisure. Almost all women were excluded as a matter of course, even though there was not yet a widespread consensual design to exclude them. The (strongly read) result is that there is no basis for an intrinsic connection between early empiricism and male dominance.

And now for the anecdote about so-called "mad Marge," which not only displays Samuel Pepys in a characteristic snit but is curiously inverted relative to later modern social constructions of gender. Margaret Cavendish, duchess of Newcastle, was already well known as a learned, extravagant eccentric when she asked to visit the Royal Society in 1667. The fellows debated her visit and decided that on May 30, Boyle and Hooke would show her experiments involving colors, mixing cold liquids, dissolving meat in oil of vitriol, weighing air, the flattening of marbles, magnetism, and "a good microscope." Pepys reports that he "did not like her at all because her dress was so antick and her deportment so ordinary." She seemed to be duly impressed by the experiments, though later, perhaps unbeknownst to the fellows, she wrote critically about the the uselessness of the new science for solving social and spiritual problems.[30] Apparently the Duchess usually wore black with a lot of silver. Pepys's chagrin is illuminated if one reads his entry for the day before. He became "horrid angry" with his wife on May 29 because she dressed in black and silver. He then returned to his office and was "mad" at her when she sent word she would change but not exactly to suit him. He thereby forwent an evening's entertainment for them both, but succeeded in rectifying his accounts and writing in his diary (which was probably what he wanted to do all along).[31] Their dress perhaps notwithstanding, the Duchess was interested in science and Pepys's wife was reasonable and conciliatory. It was Pepys who behaved emotionally, made a frivolous judgment, and was overly concerned about clothes.

Nine

Hypotheses non Fingo

Inevitably those remarks will suggest that the member of a mature scientific community is, like the typical character of Orwell's *1984*, the victim of a history rewritten by the powers that be. Furthermore, that suggestion is not altogether inappropriate. There are losses as well as gains in scientific revolutions, and scientists tend to be peculiarly blind to the former.

—THOMAS S. KUHN, *The Structure of Scientific Revolutions*

There is no way to chart the course empirical science would have taken if the Baconian practices and skeptical epistemic ideals of the early Royal Society had continued. Instead, Isaac Newton's comprehensive theories led to a suspension of relevant disbelief and gave deductive science a new lease. In that sense, Newtonianism is either a paradox in seventeenth-century secular empiricism or else belongs more to the eighteenth century.

I begin with a short account of Newton's life and personality and then return to the issues of scientific realism that have already been mentioned in connection with Boyle in chapters 3 and 8. The third section addresses Newton's reimportation of theology into science and its implications for masculine scientific identity. Finally, I suggest that Newton's competitiveness against Hooke, Leibniz, Flamsteed, and Whiston added an element of personal power to scientific identity that went beyond interest in the subject matter.

Newton's Life, Work, and Personality

Isaac Newton was born on Christmas Day, 1642, and he was a weak and sickly infant.[1] His father, lord of the modest manor of Woolsthorpe, died before his birth, and upon his mother's remarriage he lived with his grandmother. He attended day school and then public school locally. His main childhood interests were mechanics, carpentry, drawing, and writing verses. He made things, such as sundials, a windmill, a water clock, and a chair with wheels. When he was fourteen, his stepfather died and he was called home from school to help manage his ancestral farm at Woolsthorpe. His heart was not in this, and he was sent to Trinity College in 1660. He enrolled as a *subsizar* at Cambridge, which meant that he had to maintain his expenses by doing menial work.

Newton purchased a prism in 1664. When Cambridge was closed due to plague from 1665–67, he returned home to Woolsthorpe and apparently then did his major work on optics and discovered the unequal refrangibility of light rays. This eventually led to his construction of an improved reflecting telescope, which he submitted to the Royal Society in 1672 after he had been elected a fellow. During the Woolsthorpe years he also did his major work in mathematics, mechanics, astronomy, motion, gravitation and fluxions (calculus).

Newton began with several ideas from mechanism: Cartesian inertia, gravity as a universal force, and centrifugal force in circular motion. He also accepted Kepler's three laws of planetary motion, which were based on astronomical observation, and worked out the mathematics that showed how the motions of the planets conformed to Kepler's laws.[2] However, Newton's success was not communicated to others before 1669. He was secretive about his work and reluctant to publish, so that to this day it is not precisely clear exactly when he did what, and to what extent he set down the details of his intuitions before they saw formal publishable form. Following John Wiston, biographers like to say that Newton proceeded by "smell" rather than "sight."

The great contribution that won Newton resounding recognition was his connection of the then-accepted inverse-square law of force to Kepler's laws of planetary motion. In 1679, he was able to confirm the assumption that gravity was a universal force, through the proportional equality, in the same amount of time, of the distance traveled by the moon to the distance traveled by an object near the surface of the earth. In 1685, he capped this work by confirming his earlier hunch that spherical bodies exert forces on one another as though all of their particles were located in their centers.[3]

After returning to Cambridge in 1667, Newton became a Master of Arts and Sciences and, in 1668, a Fellow of the University. In 1669, Isaac Barrow backed him as successor to his position as Lucasian Professor of Mathematics. Except for three or four weeks a year, Newton spent the next twenty-six years in Cambridge, lecturing on optics and elementary mathematics.

Whether it was because of Newton's intellectual stature or, more likely, because his behavior was disconcerting even in the nonpsychologistic milieu of seventeenth-century English natural philosophy, his quirks stood out. For example, he had such an extreme reluctance to making his scientific discoveries public, because he disliked the ensuing controversy, that in 1684 the Royal Society appointed a committee, led by Edmund Halley, to remind him of his commitment to publish. The *Principia Mathematica* was published in 1687, after Halley prevailed upon him to include the third book, containing the application of his system, which Newton wanted to suppress when he heard that Robert Hooke claimed to have had the entire system before him. The book was delivered to the society by Dr. Vincent, the husband of Miss Storey, a friend with whom Newton lodged in his teens who gave an account of their relationship that has led biographers to identify her as the sole romantic interest in Newton's entire life.

It is generally related that Newton had a psychological breakdown in 1692–93, at the time of his unsuccessful attempts to secure a prestigious and profitable government post through the efforts of his friend Charles Halifax, later Lord Montagu. He wrote to Samuel Pepys that he was "extremely troubled at the embroilment" he was in and would have to withdraw from Pepys and all of his other friends. He then wrote John Locke an apology for "being of the opinion that you endeavoured to embroil me with women." Locke's response was kind and reassuring, and Newton apologized further, pleading overwork and lack of sleep. They continued as friends until Locke's death.[4]

Newton became warden of the Mint in 1695. He efficiently administered the complicated project of replacing all of the clipped silver coinage in England. This job called for bureaucratic and financial expertise, as well as technical efficiency in increasing the production of coins; he also had to acquit himself of false charges of wrongdoing. In 1699 he became master of the Mint and in 1703 was elected president of the Royal Society. He held both posts with great distinction until his death in 1727.

During his administrative years, Newton resided in London and his niece, Catherine Barton, kept house for him. Biographers have inconclusively linked her to Charles Montagu as a friend, mistress, or secret wife.[5] After Montagu's death, she married John Conduitt, who became master of the Mint after Newton's death.

Although Newton rose in fame and personal power as a scientist during the early eighteenth century, after 1695 he did not, to use his own words, make any further discoveries about the "litigious lady" of natural philosophy, or have "another pull at the moon." He was often so exasperated by the discussion his work occasioned that the pursuit of natural philosophy seems not to have been worthwhile to him. He accumulated a respectable fortune in London; lived convivially with his niece, both before and after her marriage to Conduitt; was generous to his other friends and relatives; and prevailed over Liebniz in the matter of credit for the differential calculus. He also pursued considerable scholarship in biblical prophecy and a still-mysterious career in alchemy. His *Opticks* was published in 1704 and the second edition of the *Principia* appeared in 1713. The nineteenth-century historian Augustus De Morgan remarks that while Leibniz and the Bernoulli brothers continued to advance the calculus during this period, "Had Newton remained at his post, coining nothing but ideas, the mathematical science might have gained a century of advance."[6]

Newton's Realism and Deductive Empiricism

Several points made about realism in chapters 3 and 8 are relevant again here: In Galileo's time, Catholic Church officials would accept the Copernican theory only if it were offered as an explanation of astronomical appearances, and not as a description of real celestial motions. Gassendi developed a general philosophy of science that restricted science to appearances. Boyle was fully aware of the hypothetical nature of atoms and thought that atoms could be used to explain and predict the behavior of observable objects. For Boyle, this explanatory power rested on attributing to atoms the qualities of objects that could be observed. Newton went a step further and stated that atoms would be observable as soon as a sufficiently powerful microscope was developed.

Newton's *Hypotheses non fingo* is commonly translated, "I frame no hypotheses." In light of his postulation of gravity as a universal force that acted over distances and his frequent mathematical formulations of empirical generalizations that had not been confirmed by observation, this claim is sufficiently baffling to have supported centuries of scholarly interpretation. While that work cannot be examined in detail here, several key contemporary analyses show how Newton's realism and deductive empiricism marked a revolutionary departure from the skeptical, Baconian inductivism of the early decades of the Royal Society.

Bas van Fraassen points out that in the *Principia* Newton distinguished between observed appearances and reality by stipulating that the "true motions of particular bodies" could be determined from "apparent motions" if the apparent motions are assumed to be the "differences of true motions." For Newton, real bodies exist in absolute space, in which they have absolute motion; and the center of gravity of the solar system is at rest in absolute space. Van Fraassen extracts a 'model' from Newton's theory, within which there would be 'structures' that express apparent motions mathematically.[7] While such a model makes it possible to consider both absolute and relative space and motion, it does not seem plausible that Newton would have agreed that his theory was not a representation of (absolute) reality.

The traditional interpretation of Newton's *Hypotheses non fingo* is the Baconian Royal Society caveat against metaphysical or occult speculation: Hypotheses that went beyond what had been observed and were not confirmed or otherwise supported by experience were simply not to be advanced. Philosophers still impute this strict empiricism to Newton when they are not looking too deeply into what is meant by 'going beyond' or 'being supported by' observation or experience. For example, in the entry on Newton in the 1967 edition of the *Encyclopedia of Philosophy,* Dudley Shapere writes, "Newton was perfectly clear on this point [of not framing hypotheses]," and he then supports his claim about Newton's clarity with Newton's own words: "To us it is enough that gravity does really exist, and act according to the laws which we have explained, and abundantly serves to account for all the motions of the celestial bodies, and of our sea."[8] However, this was Newton's answer to Leibniz's accusation that the theory of gravity as a universal force was inadequate because Newton offered no causal explanation of gravity. Newton's "Rule I" in *Rules of Reasoning and Philosophy* is: "We are to admit no more causes of natural things than such as are both true and sufficient to explain their appearances."[9] Since gravity is a force postulated as acting at a distance in a way that Newton could not explain, it is odd that he did not postulate the existence of gravity itself as a 'cause' that is not "both true and sufficient." Instead, he proclaimed it "enough" that gravity exists, which not only begs Leibniz's question but slips out of the methodology of early empiricism which he had previously endorsed.

Although Newton's theoretical work was presented as based on observation, it was not systematically confirmed by observation. Thomas Kuhn shows how Newton's theories did not have the experimental or observational base they were presented and accepted as having. For example, although Newton's laws of motion were supposed to be based on observation, at the time his theory was first accepted only his third law, of the equality of action and reaction, could be tested, and the experiments were very restricted.[10] Kuhn also notes that even in the *Opticks* and its "Queries," which are generally viewed as more "Baconian" than the *Principia,* Newton's theoretical conclusions went beyond the existing empirical base, and Newton carefully chose experiments that would illustrate his theories.[11] Paul Feyerabend points out that where Newton's theory of colors was inconsistent with mirror images, due to its attribution of small lateral extensions to light rays that would have required mirrors to behave like walls, he simply added an ad hoc hypothesis.[12] As a general characterization of Newton's lack of empirical confirmation for theoretical generalizations, or his tendency to generalize based on limited observation, Maurice Mandelbaum reads Newton as using *transdiction* in his descriptions of "objects or events which not only have not yet been observed but which cannot in principle be observed."[13]

Most tellingly, I. Bernard Cohen reveals that Newton did not include the slogan *Hypothesis non fingo* in the first edition of the *Principia,* but in the second edition of 1713. In the first edition, Newton began Book Three with a set of hypotheses that were changed to *Regulae Philosophandi* and *Phaenomena* in subsequent editions. In the 1690s, he even considered including a conclusion to his *Opticks* that would consist of hypotheses. Based on manuscript evidence, Cohen points out that as late as 1692 Newton was considering including a Cartesian and Aristotelian hypothesis, with which he disagreed, in the second edition of the *Principia.*[14]

However, by the late 1690s, in descriptions of his own work, Newton no longer tolerated hypotheses. It is difficult to view this change as anything other than change in rhetoric for the sake of propaganda about his own work. The hypothetical views he himself had previously put forth had not been confirmed in the interim, nor did he abandon them. Consider, for example, "Hypothesis I" of Book III of the *Principia,* which remained in all editions under that title: "That the center of the system of the world is immovable. This is acknowledged by all, while some contend that the earth, others that the sun, is fixed in that centre."[15] Newton explained to Roger Cotes, who was preparing the second edition for publication, that by 'hypothesis' he meant any proposition that is "neither a phaenomenon nor deduced from phaenomena."[16] But as it stands, this definition of 'hypothesis' does not indicate whether a hypothesis is inherently speculative, probably true, or, in some way, certainly true.

Mandelbaum argues that the correct translation of the *Hypotheses non fingo* is not "I frame no hypotheses," but "I feign no hypotheses." Mandelbaum also interprets Newton's explanation to Cotes as an intention to refer to the absence of empirical justification by the term 'hypothesis'.[17] Obviously, Newton believed that gravity, his laws of motion, the corpuscular theory, and most of his work in natural philosophy were derived from experience, though he did not formulate the rules of such derivation and show how he had conformed to them. But his claim of absolute motion in terms of a center of the universe at rest was not "derived from phaenomena" but from his beliefs about God's role in the mechanical system.

Because God is real for Newton, His literal presence in the mechanical system ensures the reality of the system. And if the system itself is real, even though parts of it, such as its center, may be unobservable, then all descriptions of parts of the system could thereby lose their hypothetical nature, insofar as they are derived from descriptions of the whole system. This construction is similar to Descartes' derivation (discussed in chapter 2), whereby the fact that God is no deceiver guarantees the truth of the sciences insofar as they are derived from a priori mathematical principles known through clear and distinct ideas. But Newton went beyond Descartes' use of God to guarantee knowledge; Newton used Him to guarantee reality. By contrast, for an atheist or agnostic, the postulation of God's presence would itself be fictitious or hypothetical and would cast serious doubt on the reality of the Newtonian system and undermine the certainty of knowledge about (and perhaps within) a Cartesian scientific system.

Newton and God

Insofar as scientific pursuits were industrious, and nature, the subject of study, was a "second Bible," God was an external presence in seventeenth-century English science. But for Newton, in what was clearly a departure from the spirit of the invisible college and the Royal Society as written by Thomas Sprat, God is internal to the scientific world view. Newton is perfectly straightforward about this in the "General Scholium" of the *Principia* (second edition), the *Opticks,* and his letters to Robert Bentley. The resulting cosmology has received a uniform exposition from traditional scholars, all of whom recognize the active role God plays in a Newtonian universe governed by mechanical forces: He was the first cause of the entire celestial system; He keeps the stars and planets from crashing into one another; He constitutes absolute space and time; He corrects for irregularities in the motions of planets and comets, which if left unchecked could undermine the harmony of the whole. I will go through these divine duties in turn and then focus on the importance, in terms of identity, of Newton's insistence that God is not the soul of the world but the Lord of all dominion.

In the "General Scholium," Newton first introduces God after noting that in the absence of air resistance, planets and comets continue in their orbits according to the laws of gravity alone. Given the complexity of the planetary system it is inconceivable that mechanical causes alone could have given rise to the patterns of planetary movement and to gravity. Therefore, he says, "This most beautiful system of the sun, planets and comets, could only proceed from the counsel and dominion of an intelligent and powerful Being."[18] In his letter to Dr. Bentley when the latter received tenure of the Boyle lectureship in 1692, Newton developed this point further, adding the assumption that God devised the sun to conveniently emit light and heat. Newton later explained to Bentley that God originally placed the large fixed stars at immense distances from one another so that they would not, due to forces of gravity, fall in on one other, and that God continually maintains these large degrees of separation.[19]

God is the foundation of space and time according to Newton. Space and time are coexistent with God because He is infinite and eternal and therefore cannot not *be* at any circumscribed time, nor can He *be* nowhere. Absolute space is literally the sensorium of God: His awareness of objects in their places is what makes it possible for them to have places in absolute space. Space itself is the same everywhere and this is what makes it possible for nature to be uniform, that is, for the same laws or regularities to apply in the heavens as they do on earth.[20] If Newton had known about Hume's problem of induction (that there is no necessity that causal connections which have held in the past will hold in the future), a similar postulation of the sameness of time could have provided necessity that the laws of nature observed in the past would hold in the future.

As the Master Craftsman He was, according to Newton, God continually preserved the perfection of the system of the universe. In the last query of the *Opticks,* Newton refers to an inevitable degeneration in the order of the planetary system, due to irregularities in the movements of comets and increases in the mass of the planets, which are corrected by God's occasional intervention.[21]

It is not surprising, in view of what God makes happen according to Newton, that He is not merely the immaterial soul of the world, existing outside of nature, but the practical and effective ruler of nature. And for that reason, God is literally a subject of science.

> This Being governs all things, not as the soul of the world, but as Lord overall; and on account of his dominion he is wont to be called *Lord God pantokráton,* or *Universal Ruler;* for *God* is a relative word, and has a respect to servants; and *Deity* is the dominion of God not over his own body, as those imagine who fancy God to be the soul of the world, but over servants. The Supreme God is a Being, eternal, infinite, absolutely perfect; but a being, however perfect, without dominion, cannot be said to be Lord God. . . . *And thus much concerning God; to discourse of whom from the appearances of things, does certainly belong to Natural Philosophy.*[22] (Last italics mine)

Newton's concept of dominion seems similar to the secular idea of ownership and propriety so important to Locke, as discussed in chapter 5. While modern psychoanalytic theory posits that ideas of God are derived from ideas of the fathers of their authors, I think it sufficient, based on skepticism about psychoanalytic theory as well as theism, to view ideas of God as projections of ideas of men.[23] That is, if God doesn't exist and psychoanalytic theory is not applied, then the question is not what kind of a man was a particular theologian's father, but more generally, What kind of a man is God? For Newton, God is a projection of a male human type who, in ordinary life, had real power over a part of the earth and its inhabitants. God was, in short, a man with authority who could be relied upon to keep what he had made running smoothly by solving problems before they arose. He was a diligent, serious, master mechanic who made excellent things that lasted. There could be no disagreement with this type of man, due to his high degree of skill and lordship over others. He had no personal attributes beyond his skill and authority.

Newton's attitude toward his own work had this kind of solemn engagement. On his own authority, he guarded his work so uncompromisingly that with his success there was a change in scientific identity, toward a sovereign, authoritative knower who was as concerned with his position in the world of knowledge, that is, with his dominion, as he may have been with knowledge itself.

Power and Paradigm

Feminist criticism of seventeenth-century natural philosophy first raises the issue of identity in the claim that science then became "masculinized." At the same time, the feminist critique exposes the unethical side of science masculinized, by analyzing how it became harmful to women and other "others." Newton's own scientific identity, in ways related to his projection of human authority onto God, and of God's divinity back onto himself, was directly harmful to his professional colleagues at key points in the promulgation of his ideas.

The kindest words about Newton the person still suggest a certain distance from others: Conduitt, who lived with Newton after marrying his niece, said that he was generous with money and entertained handsomely.[24] Newtonian biographers and scholars can usually find small ground for disagreeing with the observation of John Wiston, who worked with Newton for years before being repudiated for disagreeing with him, that Newton was "of the most fearful, cautious, and suspicious temper that I ever knew."[25] But even though the genius-terrible stories have been endlessly retold, there seems to be a consensus in the traditional literature that Newton's unethical professional behavior should be overlooked because it falls into the category of personal eccentricity. This consensus overlooks an important change in the professional ethos of science with Newton. The Baconian movement upheld cooperation as a scientific virtue, for practical as well as moral reasons, because the new fields of inquiry were interrelated in theoretical construction and shared data. Scientists needed to have access to more work than their own in order to progress; and they often addressed the same problems in research because there were intellectually popular new ideas in constant discussion, such as the Copernican theory, corpuscularianism, and gravity as a universal force.

Despite this climate of open discussion, which was the founding ideology of the Royal Society, Newton distinctively refused to share the results of his work with others involved in the same problems. He was not only reluctant to publish his own ideas but loathe to recognize publicly others of comparable stature. His first reaction to disagreement was to withdraw from natural philosophy entirely, but if he could crush an opponent, he seems not to have hesitated to do so. Newton's dealings with Hooke, Flamsteed, and particularly Leibniz support this assessment.

Newton's reaction to Hooke, early in his public career, signaled his high-handed style. When Newton delivered his account of his discoveries concerning the decomposition of light to the Royal Society in 1672, Robert Hooke and others disagreed with his explanation of his research on the colors of thin plates. Newton refused to discuss the matter, and he refused to publish his work on this subject during Hooke's lifetime, so that much of it did not appear in print before the *Opticks* in 1704. Hooke was an accomplished researcher and theorist himself, and Newton's ongoing dispute with him was exacerbated by Hooke's irritability as well as the instigation of Henry Oldenburg, who was an enemy of Hooke. At the time of the original disagreement, Newton said, "I intend to be no further solicitous about matters of philosophy." He believed himself "persecuted with discussions" and blamed his "own imprudence for parting with so substantial a blessing as my quiet to run after a shadow."[26]

Newton was more aggressive toward John Flamsteed. Flamsteed, who was Astronomer Royal, was to have his observations printed through a committee headed by Newton. Without notice to Flamsteed, and in violation of a previous agreement that required Flamsteed's consent before his catalogue of the stars could be printed, Newton printed the catalogue. Flamsteed protested angrily, and when the second edition of the *Principia* appeared Newton deleted all of the acknowledgments to Flamsteed that had appeared in the first edition. Newton was extensively dependent on Flamsteed's astronomical data, both before the first edition and in the years leading up to the second edition. Even before their disagreement over Flamsteed's catalogue, Newton accused Flamsteed of *"trifling away my time* when I should be about the king's business," because Flamsteed had caused a published reference to Newton's continuation of his research on the moon in 1699.[27]

The imbroglio with Leibniz, concerning the invention of *fluxions* or the differential calculus, shows Newton at his worst. The differential calculus was essential for computing gradual changes of direction in the curved orbits of planets. When Leibniz was in England in 1673, he heard about Newton's work on fluxions and wanted to know more. He wrote to Newton who replied through Oldenburg in 1676. But Newton wrote about the binomial theorem and sent Leibniz no more than a cipher containing the letters of the following sentence in Latin: "Given equation anywhatsoever, flowing quantities involving, fluxions to find, and *vice versa.*" The cipher was as follows: "aaaaaa cc d ae eeeeeeeeeeeee ff iiiiiii lll nnnnnnnnn oooo qqqq rr ssss ttttttttt vvvvvvvvvvvv x."[28]

Leibniz then invented a differential calculus on his own, which he communicated to Oldenburg, and in 1684 he published his method. In the first edition of the *Principia* in 1687, Newton, in a scholium, mentioned his own earlier work on fluxions and Leibniz's independent invention, recounting how he had answered him in cipher. The importance of the differential calculus soon became widely known due to further work in Europe. By 1695, Newton's followers began to accuse Leibniz of plagiarism. In 1711, Leibniz complained to the Royal Society. A committee was appointed to investigate and to defend Newton. Their findings, published as *Commercium Epistolicum* in 1712 and 1725, distorted the chronology of Leibniz's correspondence with Newton to Newton's advantage, with the result that Newton was upheld as having invented the differential calculus and Leibniz was found to have plagiarized him. Newton is generally recognized as having had an active role in this investigation—he was president of the Royal Society at the time.

When Leibniz complained that he had been condemned unheard, Newton agreed to write to him through a mutual friend, but he merely received Leibniz's letters and made notes in the margins. Leibniz died in 1716 and in the 1726 edition of the *Principia* Newton omitted the scholium acknowledging Leibniz's work on the calculus. In all subsequent comment on the subject, he either indirectly implied, or let stand, the claim that Leibniz had plagiarized his work. Most scholars today believe that Leibniz and Newton invented the differential calculus independently, though Newton did get from Leibniz the idea of an organized and permanent language of notation.[29]

If Newton's drive for power seems at odds with his early reluctance to publish, his willingness to publish when he could recast his colleagues' work to his own glory shows that he liked fame on his own terms well enough. Newton's petulance in reaction to Hooke, his harshness to Flamsteed, and his secretiveness and dishonesty concerning Leibniz introduced a competitive dynamic into scientific identity that in principle crushed the cooperative spirit of the early Royal Society. The effect of this competition can be read as a model of an ideal knower who not only discovers the truth about the universe, independently, but does so before anyone else. Following Newton, such a knower was not to be disagreed with, and intellectual competitors and opponents could be subdued through unfair, indirect, and damaging tactics, by making use of the winning knower's support and power within the institutional structure of the Royal Society.

The competitive dynamic exemplified by Newton is the antithesis of open-mindedness or tolerance of different opinions. It is amazingly consistent with Kuhn's paradigmatic model of a mature science. The resulting damage to all who are excluded from the practice of mature science and the restriction of subject matter by the leaders of the field seem to follow from the unprincipled competitive dynamic. When "winning" by those in power is supported and then lauded through institutional structures, knowledge does indeed become power—in an oppressive way.

Kuhn describes the almost exclusive use of textbooks in contemporary mature sciences, up through the doctoral level. This makes further historical sense in terms of identity theory. Those in control of a current field of knowledge have textbooks rewritten so it appears as though all previous work in their field, if it had any significance, was no more than a primitive stage of information "leading up to" their knowledge. When Newton revised his account of Leibniz's work on fluxions, he in effect tried to eject that work from the history of mathematical mechanics, so that all significant events in that history could be presented as leading up to Newton.

The question in the last section, What kind of a man is God?, now changes to What kind of a God is man? In a paradigmatic and competitive universe of knowledge, a successful scientist is an intransigent and jealous god, concerned neither with justice nor mercy, but only with having things work in the ways of his own devising. The role for lesser knowers is to help execute his program with due respect for his power and recognized authority, that is, for his dominion. That happened with Newton's name and achievements for about two hundred years. The result was that in science, as in political theory in seventeenth-century Protestant England, God acquired human attributes and men became divine.

Part Three

The Unidentified

Ten

Abuses and Uses of Children

When my mother died I was very young,
And my father sold me while yet my tongue
Could scarcely cry "'weep! 'weep! 'weep!"
So your chimneys I sweep, & in soot I sleep.
— WILLIAM BLAKE, "The Chimney Sweep"

Here a pretty baby lies
Sung asleep with Lullabies:
Pray be silent and not stir
Th' easy earth that covers her.
— ROBERT HERRICK, "Epitaph upon a Child That Died"

In Parts I and II, I have been considering the identities of seventeenth-century bachelors of science in terms of their interests in science and their positions in society. They were a new group who directed their energies toward self-invention as members of that group, at the same time that they "masculinized science." The direct subjects of contemporary feminist, emancipatory, and ecological critiques were not members of this new group, nor did they have comparable standing in society. Also, the modern identities of children, women, non-Europeans, and natural beings had not yet been constructed in the seventeenth century. Therefore, as excluded and not yet conceptualized in their own rights—albeit in ways that would often do no more than rationalize their subordination—these subjects were unidentified. In Part III I explore the seventeenth-century circumstances of and ideas about those subordinate subjects and nonpersons, beginning with children.

The above quote from William Blake sentimentally refers to the condition of young children who were sold to labor to an early death as chimney sweeps in England. The word "'weep" in the poem was meant to convey the childish pronunciation of the "sweep" that the young worker would have called through the streets.[1] Such dire conditions continued for children among the poor until the end of the eighteenth century.[2] By contrast, the seventeenth-century epitaph by Robert Herrick depicts regret at the death of a cared-for middle-class child, and it reflects the late seventeenth-century change toward kindness and relative permissiveness in child-rearing within that class.

In general, the conditions of childhood in the early modern period depended on social and economic class. This relationship is most evident in John Locke's *Some Thoughts Concerning Education,* which, through the efforts of Pierre Coste, Locke's translator, was a resounding success in eighteenth-century France and throughout Europe, as well as in England.[3]

The modern middle class was new in the late seventeenth century, and this chapter is meant to show how Locke provided that class with a blueprint for reproducing itself socially. Children were a means for middle-class social propagation according to Locke's educational principles, and he had no concept of the child as a self-motivated person who could and should develop independently of parentally imposed values. Still, historically, Locke contributed substantially to the improvement of conditions for all children, and I begin with that general context of his work.

Early Modern Childhood

The Reformation and Counter-Reformation increased the attention adults paid to children during the sixteenth century. Philippe Ariès suggests that during the medieval period, children were fully integrated into adult society from the age of six or seven, and that outside of monasteries and universities little attention was paid to education. When increased concern with religious belief and more specialized economic structures throughout society made it necessary to "prepare" children for adulthood, wide-scale public education took hold in Europe and England.[4]

The early phases of this new focus on childhood, through education, were not nurturing. In schools and places of apprenticeship, corporal punishment increased in the sixteenth and seventeenth centuries, and all young people became subject to more intense scrutiny and control. As an object to be educated, the child was viewed as obstinate and sinful. At home and in schools, it was believed essential to "break the will" of children so that they could acquire knowledge and virtue. Teachers reacted brutally to academic as well as moral lapses. Thomas Tusser's 1560 lines are often quoted on this point:

> From Poules I went, to Aeton sent
> To learne straightwayes the Latin phraise
> Where fiftie three stripes given to mee at once I had[5]

Whipping the buttocks with birch sticks until blood ran and striking the hands and face with wooden fistulas that raised blisters became routine pedagogical practices. Corporal punishment was extended to adolescents and displaced the medieval custom of imposing fines; until 1660 in England, even college students were subject to whippings. Adults were also commonly subject to whipping and placement in stocks during the sixteenth and seventeenth centuries. However, adults of higher rank could usually avoid such treatment, while it was administered democratically to children.

Conditions of childhood were abusive in other ways not directly concerned with education. Parents who had the means sent their infants away from home to be nursed by poor women, in unhygienic environments, until they were toddlers. This practice is cited by contemporary historians as a major cause of the child mortality rate of one-quarter to one-third. Infants were also immobilized by swaddling during that stage of life: Their bodies were wrapped in strips of cloth to form solid mummy-like packages, which could be hung on hooks for the convenience of caretakers. After infancy, children of all classes were subject to frequent physical abuse by parents or tutors.

Child-parent relationships were distant during this time. Children were expected to address their parents formally, in terms of great respect. This carried into the widespread requirement that adult children always kneel or stand in the presence of both their parents. Parents, usually fathers, also chose or rejected marriage partners until the late seventeenth century among the middle class, and well into the nineteenth century among the aristocracy. These decisions were made on the basis of property and status; among the poor, though young people had more freedom to choose marriage partners they often did so with similar calculation. Parental choice of career or career-type, linked to property interests and social class, seems to have been the most enduring prerogative.

Until the end of the seventeenth century, there was no adult respect for the sexual innocence of childhood. Children were neither shielded from the nakedness or sexual behavior of adults nor protected from adult interest in their own nakedness and sexuality. For example, when Louis XIII was a toddler, his relatives and servants continually teased him about his penis and played sexual games with him. He and his sister were placed naked in bed with the king, "where they kissed and twittered and gave great amusement to the king."[6] Homosexuality and sadism were common throughout educational systems, particularly in boarding schools for boys, and parental and clerical indignation did not lead to reform until well into the eighteenth century. Nonetheless, many were beginning to object to excessive corporal punishment as a pedagogical tool. John Aubrey wrote of Richard Busby, headmaster at Westminster while Locke was there: "Dr. Busby hath made a number of good Scholars, but I have heard several of his Scholars affirme, that he hath marred a number by his severity more than he hath made."[7]

Locke's *Thoughts Concerning Education*

Locke never married, and there is no evidence that he ever had an illegitimate child or was in a relationship that might have led to one. Nonetheless, his extensive experience with young people as a doctor, Oxford don, and private tutor qualified him as an authority on education. As Lord Shaftesbury's family physician and confidant, Locke arranged his son's marriage and supervised the upbringing of his grandson; he also took Caleb Banks, the son of a friend of Shaftesbury, on the grand tour. Many of his charges in later years expressed gratitude for his tutelage and seemed to have liked and respected him.[8]

It was therefore appropriate that Edward Clarke, a distant relative and friend, ask Locke for advice on how to raise his heir, Edward Jr., who was eight at the time. Locke first wrote to Clarke in 1683 when he was exiled in Holland after Shaftesbury's death. The letters went through several revisions before they were published anonymously in 1693 and then evolved into *Some Thoughts Concerning Education,* which went through twenty-five editions by 1800, five of which Locke supervised before his death in 1704.[9]

From the beginning of his correspondence with Clarke, Locke kept his advice general and allowed for strategies to deal with individual variations in temperament. For example, he told Clarke it was necessary to observe his son's "temper whether bold or timorous, careless or curious, steady or unconstant, friendly or churlish, etc., for your having once established your authority and got the ascendant over him, the next thing must be to bend the crooks the other way if he have any in him, and apply proper methods to his peculiar inclinations."[10] Locke also wrote Mrs. Clarke about the upbringing of her daughter Elizabeth, for whom he showed particular concern and affection throughout her childhood. Girls were to be raised according to the same principles as boys, with slight concessions made to their being the "softer sex."[11]

Many other parents followed Locke's advice, not only because he was John Locke of the *Essay,* before the *Education* was published, but because this advice seemed reasonable and likely to bring desirable results. The generality of the advice must have made it useful over a broad range of family conditions, although it was applicable in an enduring way only to the group that was to become middle class. The following interpretation of Locke's contribution to education is meant to depart from the current view that Locke was reflecting middle-class values in that work. Rather, Locke was *creating* those values for family life. Families are social structures, and the modern middle class family, beyond biological reproduction, is no more "natural" than any other form of organization. For the middle-class nuclear family form to survive, either those who are born into it must learn how it works and recreate it when they grow up, or new people need to be recruited into that lifestyle. Obviously both kinds of perpetuation have taken place during the modern period. Locke's *Thoughts Concerning Education* addressed the inheritance, through parental training or 'breeding,' of middle-class nuclear family mores.

The political importance of Locke's distinction of parental from political authority in the *Second Treatise* has already been discussed in chapters 1 and 7. As part of this distinction, Locke argued that children do not automatically inherit their citizenship, but rather choose it. The social result of Locke's distinction between parental and political authority was to enforce private parental autonomy in organizing the lives of children. This can be read as an enforcement of liberal popularism or of patriarchal power within the family, or both.[12] In either case, it ironically restricts the liberty of children before they are old enough to choose their citizenship or religion; and since their education would be broadly determined by the religion and civic status of their fathers, it would predetermine their choice of citizenship and religion as well.

Locke was specifically interested in raising male children as *gentlemen,* or men of property and affairs. The justification for his interest was that if these boys grew up to be capable of conducting their own business and public affairs, in a certain way, everyone would benefit, "For if those of that Rank are by their Education once set right they will quickly bring all the rest into Order."[13] Locke also had a clear idea of what he called "a Gentleman's Calling," which was "to have the Knowledge of a Man of Business, a Carriage suitable to his Rank, and to be Emminent and Useful in his Country according to his Station."[14]

Such gentlemen had property and tenants, were heads of families, and might have been local magistrates and sheriffs, as well as members of Parliament. They made up 4 to 5 percent of the population of England then and were the effective ruling class.[15] It may be a misnomer to refer to them as the middle class, but the term is accurate in the sense that they were not royalty, all did not have hereditary titles, and many did not have great wealth. Elite though they were, they were engaged in ordinary practical life and expected to be active citizens.

Locke begins the *Education* with an endorsement of Juvenal's *mens sana in corpore sano,* presented in English, and proceeds to lay out a very specific, austere regimen of diet and personal habit for children: a lot of bread, little meat, restricted fruit, trained bowels, hard beds, early rising and exercise outdoors with bare heads and wet feet, all year round. In general, the child was to be exposed to a variety of discomforts so as not to be disconcerted or made ill by similar inconveniences in adult life.[16]

In these early sections of the *Education,* Locke seems to start from scratch in his advice to parents, as though everything relating to child-rearing needs to be approached with no prior preconceptions, traditions, or personal preferences. He is wary of the "fondness" of mothers because of its weakening effect and believes that doctors and medicine should be avoided as much as possible. Throughout the *Education,* he warns that children ought to be kept away from the corrupting influence of servants.[17] Children were to be in the company of their parents as much as possible, and Locke frequently gives examples of mothers teaching young children a variety of subjects, including Latin.[18] All rewards and "good things" were to be dispensed directly by parents, though when punishment was necessary it could be administered by servants.[19]

There is no sign that Locke expected the parents themselves to have been raised according to his advice, nor were they expected to acquire the Spartan habits to be imposed on their children. That contrast in the first two Lockean generations would presumably create heirs who would be hardier than their parents, because, as Locke tells us at the outset, "of all the Men we meet with, Nine Parts of Ten are what they are, Good or Evil, useful or not, by their Education."[20]

Thus, children returned to a household after weaning are first set off from everyone else by a disciplined regimen. And, as Locke's advice develops, it becomes clear that the bodily program is not only healthful, based on his experience as a physician, but also a necessary foundation for his program of the mind (based presumably on his philosophical and social wisdom).

> As the Strength of the Body lies chiefly in being able to endure Hardships, so also does that of the Mind. And the great Principle and Foundation of all Vertue and Worth, is placed in this, That a Man is able to *deny himself* his own Desires, cross his own Inclinations, and purely follow what Reason directs as best, tho' the Appetite lean the other way.[21]

Locke assumes that if children are not taught this kind of self-discipline, they will not have it as adults: "He that is not used to submit his Will to the reason of others, *when* he is *Young,* will scarce hearken or submit to his own Reason, when he is of an Age to make use of it."[22] Thus, for Locke maturity was not the result of a natural process, but rather of the right education. Gentlemen's children should be used to submitting their desires to their parents' wishes and going without objects of longing, "*even from their very cradles.*" Wanting something is never a reason for getting it if it is not otherwise "fit," and if the child protests or shows too much eagerness for an object, denial should be even more resolute.[23]

When Locke's gentlemen's sons are young, they should be subjugated to their parents, though they can be friends with them when they are older. Although young children should be in awe of their parents, great severity is not called for. Corporal punishment not only may break the will and cause "dejected minds" and "low spirits," but it also teaches the child to avoid pain rather than be ashamed of the wrong action being punished. Children should develop fortitude, or an indifference to pain, and sometimes, if it is done in a kindly way, physical pain should be deliberately inflicted on a child for this purpose. Neither should children be rewarded so that they think that happiness lies in pleasures of food, clothing, amusement, and the like. Rather, they should be happy because they have their parents' approval. The appropriate punishments or rewards should flow from the parents' disapproval or approval. While praise should be publicly proclaimed, rebukes ought to be privately communicated to the child so that he will always have something to lose in terms of his reputation—if the child thought his good reputation were lost, there would be little reason for him to be virtuous.[24]

It is interesting that although Locke frequently claims that the goal of education is virtue, he does not suggest that children be taught to find happiness in being virtuous. That their happiness is to lie in their parents' approval and their own good "reputation" suggests either that Locke did not believe that children could have any real understanding of moral virtue or that he wanted to set the parent-child relationship up so as always to give the parent the maximum amount of control. A child who understood moral values might not always defer to parents, but a child who was dependent on parental esteem for well-being would. The result is that Locke posits the child as a passive recipient of its moral education. We don't know at what age Locke would have thought a child able to judge the moral worth of its actions, independently of parental opinion. Indeed, the irrelevance of such a question in his scheme of parent-child relationships suggests that he was mainly interested in virtue insofar as it was useful to support adult authority. Locke's insistence on getting children to practice what is right, as opposed to teaching them rules, while highly praised by educators in subsequent generations, enforces a lack of independent reasoning about moral issues.[25] Although Locke did not endorse beating children for most offenses, he did think that obstinacy and rebelliousness had to be countered with corporal punishment.[26] This exception reinforces the importance he placed on making the child mentally submissive. Locke did not get far from the doctrine of "breaking the will" in education, such was his zest for control.

Locke believed that it was preferable to educate children at home, under well-selected tutors, than to send them to school where they would learn vices. He admitted, however, that more confidence and skill at "bustling" for oneself would be achieved through public education.[27] The tutor should himself be virtuous, well-bred, wise about the ways of the world, and a good judge of human character so that he would be able to assist the heir in the transition to adult society without wildness or self-indulgence.[28] The scholastic part of the education taught by the tutor ought to be minimal, general, and above all useful: "And since it cannot be hoped, he should have Time and Strength to learn all Things, most Pains should be taken about that which is most necessary; and that Principally look'd after, which will be of most and frequentest use to him in the World."[29]

Locke says that learning is the least part of education. He continually emphasizes the importance of play and recreation for children, for two reasons. First, children need play and learning itself can be introduced as a form of recreation. Second, when a child plays, his parents have an opportunity to observe his nature and pinpoint his moral defects.[30]

The child's earliest instructions should begin with languages, preferably through conversation rather than the rules of grammar.[31] Latin and Greek are not worth great pains to inculcate, as was done in public schools at that time, but astronomy, geography, anatomy, history, and geometry are worthwhile parts of the home curriculum.[32] There isn't much use in having children memorize large blocks of prose, write themes on Latin expressions, or read or write poetry.[33] Indeed, what Locke has to say about poetry directly illuminates the purpose behind almost everything else he discusses in the *Education:* "And if he have a Poetik Vein, 'tis to me the strangest thing in the World, that the Father should desire, or suffer it to be cherished, or improved. Methinks the Parents should labour to have it stifled, and suppressed, as much as may be." Locke then goes on to associate poetry with gambling, neither of which "bring any Advantage but to those who have nothing else to live on." He therefore concludes that "if you would not have him waste his Time and Estate, to divert others, and contemn the dirty Acres left him by his Ancestors, I do not think you will much Care he should be a *Poet,* or that his School-master should enter him in Versifying."[34]

That fathers have an unquestioned right to choose their sons' occupations, that poetry and imagination are opposed to practical skills in the world, and that young people's interests can be controlled to everyone's benefit—all of those assumptions stand behind the above lines. Locke also advised that the gentleman's son learn at least one manual trade, such as painting, woodworking, gardening, or metalworking.[35] And he enthusiastically recommended the study of accounting "to make him preserve the estate he has."[36]

Not everyone agrees that Locke's principles of education were biased by class interests, as I have suggested. The connections between the educational principles and Locke's more abstract writings on intellectual psychology in the *Essay Concerning Human Understanding* and the *Conduct of the Understanding* have been offered by James Axtell and other traditional scholars as evidence of the universality of the *Education.*[37] Locke's psychological generalizations about the importance for learning of the association of ideas, experience, and clear and distinct ideas do purport to be universal principles, true of all human beings. The same can be said of his emphasis on the lasting nature of what is learned early in life. His instructions to parents for effective education and socialization could not work, insofar as they are based on those principles, if the principles themselves were false. But it does not follow from this that the instructions themselves are meant to be applied universally—especially in light of Locke's own claims that he is offering thoughts about education for gentlemen and that, even within that group, differences in rank will require differences in emphasis and application.[38] Furthermore, if the knowledge of universal psychological generalizations can be used to make training effective, it does not follow from this that they should be used for any particular training, or for any training at all. It could be argued that no human being ought to be educated by manipulating psychological regularities over which he or she has no control. Such "training" interferes with (or fails to support) an individual's ability to understand her situation and make informed choices.

Traditional Locke scholars may also overlook the fact that the early modern middle-class family was a private institution centered within a house that required a full-time supervisor, or wife and mother. In principle and restricted practice, Locke did not exclude girls from home education. But because marriage and motherhood were the only socially valued occupations for middle-class women, the daughters of gentlemen could not have inherited their fathers' civic status. By the eighteenth century, the intellectual components of middle-class female education shrunk to ensure marriageability, and restrictions on physical exercise, as well as body-shaping contraptions, weakened girls physically. The resulting feminization of women worked against the inadvertent feminism in Locke's theories of education and politics (though his inadvertent feminism was important and will be discussed in the next chapter).

Liberal and humane though Locke's ideas of child-rearing may have been in a context of overt abuse, his ideas were ultimately exploitative of the child as a person. Children were to be dealt with instrumentally, for the good of a restricted segment of society, as shaped by their fathers' ambitions. What does this say about Locke and other bachelors of science who may have shared his views? As bachelors they did not have the opportunity to apply their liberal political views to personal family experience, and this may have thwarted their empathy for young people. In concluding this chapter a more empathic approach to childhood merits consideration.

Notes on Childhood, Then and Now

Laura Purdy locates Locke's blank slate theory of education as an extreme that can be opposed to Rousseau's "growth metaphor." The growth model conceptualizes childhood development as internally driven and capable of unfolding without significant external input.[39] In her assessment of the benefits of both models, Purdy observes, "What we need is to isolate those characteristics that are essential for every child to develop in order to live a satisfying and moral life, then figure out how to achieve those goals."[40] However, Purdy and many other philosophers of education fail to notice that the essential characteristics for a satisfying and moral life may be too closely tied to individual temperament to be determinable for others, even if those others are our own children. For example, Locke's dismissal of poetry as unworthy is still relevant. Even on social levels not as privileged or materially minded as Locke's landed gentry, few middle class parents today would encourage their adolescent sons to become poets (or rock musicians). But for those who choose such an occupation, in disobedience or rebellion, there may be no better path to a satisfying and moral life. For them, the "essential characteristic" might be personal integrity, which the child has to develop independently.

Following my interpretation of Locke on education, it could be argued that much of the present ideology about family life and ties, including late twentieth-century rhetoric about "family values," can be reread as directions for perpetuating the social conditions of those who speak it. This is not to say that the "real" motives behind such ideology are necessarily economic. Besides general desires for keeping and increasing their power in the world and leaving their "mark" on others, people want those others to be like them. Those who are already subordinate, such as children, are ready subjects for this kind of influence.

If children were not inherently subordinate, the liberal extreme of prescriptions for early education would not be called "permissive." In fact, few have ever considered the total emancipation of children as a serious option. When children are now accorded rights, adult advocates are appointed to represent their interests as they would presumably want them represented if they were rational adults. The justification for the subordinate status of children is that their undeveloped capabilities would be considered diminished capabilities if they were adults. Most people outgrow the diminutive capabilities of childhood; then, from adult vantage points, they are no longer part of the subordinate group of children. This structure of temporary subordination always puts adults past the position of direct concern about how children are abused and used. For example, the extraordinarily oppressive practice of swaddling in the seventeenth century is still rationalized as having been necessary, even by 'permissive' contemporary social historians.[41] It is very difficult for adults to look at the situations of childhood, in any era, from the position of the children themselves, and this is no different from other perspectives that accompany unequal distributions of power.

Almost all theories of childhood and child-rearing consider the development of children from the standpoint of adult society. The models of childhood are in this sense teleological: childhood is for the sake of adulthood, which is its final cause or goal. But another way of looking at childhood, which may be more sensitive to how it is experienced by its subjects, is to consider the culture of childhood. When children are separated from adults in groups with other children, they learn from their peers and from those slightly older. On playgrounds and in schools, this culture of childhood is ongoing because, though everyone eventually leaves, new younger children are always joining it. The main area of research on this subject seems to be childhood play and games, perhaps because play and games have thus far been the main distinguishing aspect of the culture of childhood in the modern period. Ariès notes that with the early modern formation of the middle class, previous adult practices of play and dressing-up were relegated to children.[42] Indeed, the entire culture of childhood may, ironically, be an effect of the separation of children from adults in order to mold and control them—ironic because children reassert their autonomy in often uncontrollable ways within their own culture.

The importance of focusing on peer cultures of childhood is that it opens up a perspective from the child's point of view. At any time, the continuous culture of children's groups need not have the same values as existing adult cultures. It is probably no coincidence that when adults discover a new lifestyle or religion, or, as in the seventeenth century, a new social class, their first educational innovation may be to remove their children from the culture of their peers and educate them at home. Of course, the child's peer culture may be vicious and unjust and its value system in all respects inferior to the ideals of enlightened and ambitious parents.[43] But in its defense, like any other peer culture, the child's culture allows participants to develop their unique subjectivities and individual agency. At times this amounts to no more than a struggle to conform in which those who do not succeed invent themselves through ordeals of rejection and humiliation. Nevertheless, peer relationships have the redemptive potential of keeping us honest. Our peers can unmask us and see through our dissimulations because in their similar circumstances they have constructed like personas and mastered comparable tricks.

Relationships with unequal power distributions, such as parent-child interactions in middle-class families, are often constructed around objectives that preclude such honesty. The child needs to convince the parent that she is what the parent wants her to be in order to get what she wants from the parent. And the parent tries to present herself as a model of what she wants the child to be. Their circumstances are sufficiently different so that they can fool one another about their motives and achievements. When parental controls are relaxed or children succeed in rebelling, the perpetuation of the parents' social class becomes uncertain—for better or worse—which is why heads of households who are satisfied with their lifestyles abhor "permissiveness."

Members of the middle class have to work in order to maintain their positions in society, which is probably why education is so important to them. Their children have to be taught how to do the same kind of work their parents do, both inside and outside the home. If wealth is inherited and social position unassailable, it is not as imperative to assume the narrow burdens of formal learning, though socialization may be even more important. If parents are poor, unless they are ambitious and upward mobility is available, there is less that can be lost if children are not educated in the right way. (However, the poor may have other reasons for abhoring permissiveness in child-rearing.)

The important contemporary insight from Locke's *Thoughts Concerning Education* is that the position of children in a class-structured society is not only determined by the class they are born into, but by the assumption that they will carry on the work of that class and perpetuate it socially. Whether parents of any class have a right to impose their class identities on their children through education is a question yet to be raised. If history can teach us, the lesson here is that practices of child-rearing that were seen as necessary or unobjectionable at one time may look abusive and manipulative in different historical circumstances. Although the culture of childhood has received renewed attention in recent decades, few researchers or theorists seem inclined to investigate that culture for positive answers to the type of question Clarke asked Locke, "How shall we raise our children?" Insofar as the popular youth culture of the late twentieth century falls short of adult middle-class values and aspirations, it is mainly—except for the window of social consciousness and recreation associated with the late 1960s—a subject for popular criticism. But perhaps Clarke's question might be recast as, "If we were children, how should we raise ourselves?"

Eleven

Wifemen and Feminists

wo'man I. Adult human female . . . [OE *wifmon(n), -man(n)* (WIFE, MAN), a formation peculiar to English, the ancient wd being WIFE]
—*The Concise Oxford English Dictionary*

By a woman, then, I understand an individual human being whose life is her own concern; whose worth, in my eye (worth being an entirely personal matter), is in no way advanced or detracted from by the accident of marriage.
—CICELY HAMILTON, *Marriage As a Trade*

To gain the supreme victory, it is necessary, for one thing, that by and through their natural differentiation men and women unequivocally affirm their brotherhood.
—SIMONE DE BEAUVOIR, *The Second Sex*

The *Oxford English Dictionary* suggests that the premodern English meaning of *woman* was a man's wife, or perhaps a man who was a wife. Mary Wollstonecraft in the eighteenth century, and John Stuart Mill in the nineteenth, argued for women's rights on the grounds that they would make women better wives and mothers. Cicely Hamilton and other early twentieth century feminists insisted that women ought to have identities separate from being wives. Simone de Beauvoir's classically liberal feminism rested on a generic idea of humanity—if not real human nature—that would enable women to have moral equality to men. Few contemporary feminists would feel comfortable about de Beauvoir's call for *brotherly* solidarity because its masculine gender biases any sense of generic humanity. De Beauvoir also seems to assume that there are inherent differences between men and women apart from the social roles that have historically accompanied constructions of female and male gender. That type of assumption sets limits on studies about how natural biological differences are theoretical constructions that have been projected onto men and women at different times in history.

However, biological foundationism dies hard. Even a full-fledged view of human biology as a cultural construction presupposes a physical foundation for difference between those human individuals classified as male and female. In Thomas Laquer's analysis of premodern history, gender used to be fixed as an expression of status based on social power structures and their resulting social roles, and physical sexual difference, or what we would now call biological difference, was malleable and derivative.[1] Laquer's approach is supported by the self-images of independent early modern women, such as Queen Elizabeth I and Aphra Behn, who associated their professional achievements with something male within themselves.[2] His analysis is fascinating for its potential to upend some contemporary critiques of gender and sexuality.

Nonetheless, I think that by the seventeenth century, as is the situation now, the difference between men and women was widely assumed to rest on something physical. I am therefore going to assume a physical foundation for male and female sexual difference and a cultural foundation for male and female gender difference. A less historically contingent analysis of sex and gender might proceed with a variable for the foundation of male and female difference that could be either physical or cultural. But if the foundation were cultural, there would be a problem picking out the group of women in historical periods when women were assigned different cultural identities than in our own. Laquer himself distinguishes between the culturally constructed body that is the result of cultural contingencies and the "real" body that is the subject of medicine in his own experience.[3] So long as a theorist believes in such a real physical body that is different for men and women, cultural foundationism, as an alternative to biological foundationism, is not being taken seriously.

The physical differences between women and men that were commonly assumed to obtain in the seventeenth century were revised during the eighteenth and nineteenth centuries so as to reinforce new constructions of the psychology and social role of women. The identity of women in the seventeenth century was not qualitatively different from the identity of men. Women were considered to be generally like men, but inferior to them.

This chapter begins with an examination of that negative female identity. The next section is a discussion of Hobbes' feminist account of women in the state of nature. The last section addresses what today could be considered genuine feminism in the seventeenth century, that is, critical intellectual work by women about the position of women in society.

Negative Feminine Identity

Alice Clark's 1919 historical analysis of the loss of female productivity due to the rise of male-dominated capitalism in seventeenth-century England shaped the present feminist scholarly consensus about the fall of woman in the early modern period. With the increase of wage labor during early industrialism, men began to work outside the domestic economy. Women who had previously worked alongside their husbands in trades, crafts, and farming and lost those positions, had less economic responsibility in the household. Crafts previously dominated by women, such as midwifery, brewing, and tanning came under the control of men. Extended households that previously consisted of nuclear families, relatives, servants, apprentices, and other employees shrank to nuclear-family households. In the middle classes, the social world of women began to contract to private, nuclear-family households. Female employment outside the home or beyond nuclear-family care-giving became restricted to ill-paid and physically debilitating work, primarily in agricultural labor and the cloth trades. It was virtually impossible for wage-laboring families to rise above subsistence and difficult for single women or women with dependent children to attain even that.[4]

During the eighteenth and nineteenth centuries, ideologies of romantic love, motherhood, intellectual and physical debility, and emotional instability were constructed to account for the economic and social relocation of middle-class women's lives that began in the early modern period. Significantly, the twentieth-century feminist criticism of these modern female identities has accompanied economic and social changes that have moved middle-class women back into positions of economic and social responsibility. However, it is not clear exactly "who" women were before the changes in the seventeenth century. As "wifemen," women worked at the trades of their husbands. When men began to work outside of the home in market economies, however, they were employed as single individuals. In effect, this means that previously productive wives were made redundant by the new centralization of work places. Thus, as a wife, an early modern woman did not have control over her work life because it was subsumed under her husband's work identity, and only insofar as she worked alongside him. The subsumption of a wife's work identity under her husband's meant that she could not have a work identity of her own. This redundancy of wives' work was not limited to the laboring classes. Alice Clark documents how at the beginning of the seventeenth century upper-class wives were actively engaged in household, estate, and political affairs, whereas by the period after the Restoration their granddaughters were rendered idle by the specialization of employment and increased family wealth resulting from accelerating capitalism.[5]

Unmarried mothers were punished with varying degrees of severity throughout early modern Europe. But even married women did not have strong identities as mothers. Infant and adult mortality rates were high and there was no general likelihood that infants and children would be raised by their biological mothers, especially given the use of wet nurses. Miranda Chaytor and Jane Lewis note that in defamation cases, "whore" was the most common term of abuse. Women based their moral worth as women on chastity and fidelity. There are no recorded charges or accusations of what today would be considered the abuse or neglect of children. However, next to witchcraft, infanticide was the most frequent female crime and was punishable by execution, often by drowning.[6] Although, as Aristotle had stipulated, women were believed to furnish no more than the physical material for pregnancy, the infertility of childless couples was blamed on wives. It was also believed that pregnancy could not take place without female orgasm, a belief that blocked charges of rape if pregnancy resulted.[7]

Women's identity in the seventeenth century was strongly sexual in ways that might be difficult to comprehend through subsequent (European, English, and American) middle-class lenses. While the partners of Samuel Pepys's adulterous seductions seem to have been motivated by money and the career advancement of male relatives, Pepys's sexual attention was clearly focused on their bodies, especially their breasts. At the time, fashionable women wore dresses cut low over their breasts in all seasons, and breasts were replacing the womb as a symbol of female sexuality.[8] Pepys had sexual relations with more than fifty women during the nine years recorded in his diary (1660–69), none of whom were professional prostitutes.[9] While there is no reason to assume that Pepys represented a norm, women were generally viewed as the sexual sex during the seventeenth century. It was known that they were able to have more orgasms than men, and it was believed that their desires were stronger than male sexual desires. In addition, however, men thought that female sexuality was dangerous: Sexual intercourse debilitated men by using up the common material of brain and blood that was expelled in semen; women could be a source of venereal disease, and of unwanted offspring and attendant financial obligations.[10]

The strength of seventeenth-century female sexuality seems not to have been associated with sensual aesthetic interest in the female human body either clothed or naked. Men dressed even more lavishly than women when gaudy clothes were in fashion. Women did not wear distinctive underwear and, though female "drawers" were coming into use, they were considered immodest.[11] Neither sex bathed habitually, and until late in the century among the upper classes, there was little privacy surrounding any bodily function.[12]

Female sexuality was not anchored to an idea of femaleness as fundamentally different from maleness, and the assumed underlying similarities are puzzling from present perspectives. For instance, official records did not specify the sex of a "child" when births were recorded. Young boys and girls were dressed alike as infants in swaddling clothes, and after that in "coats" (petticoats) until the age of four or five.[13] If, as doctors and social scientists now tell us, male and female gender is established by the age of three in subjective personal identity, it is not clear how, or when, or even if, it became fixed during the early modern period.

There was also an assumed underlying physical sameness about adult men and women. All bodily fluids in both sexes were considered to be the same material in different forms. Not only could menstrual blood become milk during lactation, but women with high body heat were believed capable of producing semen, and men with low bodily heat were thought able to lactate. Indeed, body heat seems to have been the one essential difference between the sexes. On Aristotle's model, the fundamental cause of women's inferiority to men was that men were warmer and dryer than women, and women were colder and damper than men.[14] The well-known Aristotelian relativity of perceptions of temperature and the corpuscularian relegation of temperature to the realm of (apparent) secondary qualities seem not to have been applied to this important sexual difference.

Andreas Vesalius, the esteemed sixteenth-century anatomist, had depicted female sex organs as inversions of male sex organs: "The organs of procreation are the same in the male and the female . . . for if you turn the scrotum, the testicles and the penis inside out you will have all the genital organs of the female."[15] It was believed that the greater warmth of the male body was responsible for pushing these organs out. European doctors frequently reported cases of women who became men through strenuous physical exercise that raised their body temperature sufficiently to push out their sex organs, although, as Merry Wiesner wryly notes, there were no reported cases of low body temperature turning men into women.[16]

The presumed physical inferiority of women carried over into reproduction. William Harvey reported, as a result of his dissections of slaughtered post-coital doe, that sperm fertilized eggs without coming in contact with them, an idea of "action at a distance" that reinforced ideas of male potency. Anton von Leuenhook reported microscopic observation of sexually differentiated *homunculi* in sperm, which meant that women did not contribute to the sex of a child.[17] The uterus was considered a problematic organ, capable of independently "wandering" in the body to cause mental illness. Menstruation was viewed as a healthful function because it discharged bad blood, which after menopause built up to cause not only physical illness but greater sexual desire.[18]

Women were generally believed to desire men sexually more than men desired women because the Aristotelian model decreed that something imperfect always desires something closer to perfection.[19] If women accepted the Aristotelian model, and if sexual desire is based on what people believe, it is conceivable that women did desire men more than men desired them during the early modern period. We know that some Victorian women behaved as though they had very little or no sexual desire, when the polarity of sexual difference in desire was reversed.

The biological foundationism of the seventeenth century was solid enough to preclude recognition of female homosexuality on the grounds that vaginal penetration did not take place. Homoeroticism between women was not taken seriously as a punishable offense in law, as it was for men. Nonetheless, cross-dressing and other violations of gender norms that upset economic and social power were punishable, in some cases by execution of the women who behaved inappropriately for their "sex."[20] This suggests that despite genital isomorphism, sexual activity was defined around males, and that gender difference was mediated by the higher social status of males.

Thus, in comparison to men, women had strong sexual identities in the seventeenth century, combined with weak work, physical, and parental identities. But their sexual identities seem mainly to have rested on assumptions about female sexual desire and masculine experiences of the effects of female sexuality on men. Aside from its necessity for procreation and the later Protestant support of sex in marriage as a source of mutual "comfort and endearment," there is little evidence that sexual attractiveness or sexual activity had any socially or morally justifiable value.[21]

There was much "bawdiness" in public behavior and popular entertainment before and after the Puritan hegemony under Cromwell. But even though the middle class did not support church or governmental regulation of private sexual behavior, by the end of the century Puritanism dominated public morals.[22] For example, when William assumed the throne in 1688, Aphra Behn, whose sexually explicit plays had enjoyed great success during the Tory Restoration, saw the hand writing on the wall and knew that she had no place in the new Whig order:

> Though I the wondrous change deplore
> That makes me useless and forlorn.[23]

However, even in Pepys's heyday, Behn's plays and verses rarely went beyond the conventional wisdom about sexuality. Her poem "The Disappointment," an account of Lysander's sexual impotence in the face of Cloris's desirability, has been interpreted as an expression of Behn's own erotic frustration with John Hoyle because he was homosexual or bisexual or simply failed to reciprocate Behn's passion for him.[24] But in Behn's poem, after Cloris has fled, blushing with "both distain and shame," Lysander ends by cursing

> The shepardess' charms
> Whose soft betwitching influence
> Had damned him to the hell of impotence[25]

Behn explains clearly that "Excess of love his love betrayed," which suggests that even at the highpoint of sexual permissiveness in London, and at the peak of their desirability, women were still blameworthy for their sexuality.

The combination of danger, one-sided lust, and inferior masculinity at the locus of female sexuality resulted in a construction by men of the otherness of women that had none of the social value of male gender, and little in its own right that was positive. Seventeenth-century men's gender seems strangely bereft of sexuality in its absence of autonomous sexual desire–desire was evoked by women and could also be stilled by them, as Behn's poem describes. Returning for a second to male bachelorhood and the lack of affect in the Lockean person as discussed in chapters 4, 5, and 6, the absence of an idea of independent male sexual desire in how men thought about themselves would seem to reinforce other reasons for celibacy and isolated individuality.

Hobbes' Latent Feminism

Locke's *Second Treatise* argument for the rights of women, based on the authority of female parents, has already been noted in chapter 7. It has also been noted that both Locke and Descartes advocated education for women and did not seem to assume an inherent intellectual inferiority of women. More direct advocacy of education for women by seventeenth-century feminists will be discussed in the next section. However, the political writings of Hobbes had an interesting potential for an early modern theoretical foundation for women's rights. And insofar as contract theory is still viable in political science and philosophy, Hobbes' feminism remains relevant.

Locke's distinction between governmental and patriarchal power influenced the demise of the patriarchal model of government in political theory by 1700. However, the patriarchal model of the family was considerably more enduring. Carole Pateman points out that within patriarchal family power there is a distinction between the rights of fathers over children and the rights of husbands over wives. Pateman conclusively argues that father-rights would need to rest on husband-rights for any political theory to be anthropologically accurate, because men must have the right to have sex with women who will conceive and bear children before they can become fathers.[26] (Pateman calls this an anthropological rather than a biological fact, presumably because she means her analysis to cover societies in which it is already known that heterosexual intercourse is the cause of pregnancy.) This means that the institutional rights of husbands over wives would require the consent of wives according to contract theory, or else that consent in Hobbes' terms could be read from the apparent willingness of wives to live under the domination of their husbands. The latter kind of implicit consent was no problem for Hobbes since he believed that even infants are parties to contracts in situations where they are nurtured.[27]

Patemen notes that in Hobbes' original state of nature, women were assumed to be as free as men and a child could not have "two masters." Because the woman originally had control of the child and women were of comparable strength to men, it would have required violence for a man to take a child away from its mother. A woman was therefore the "lord mother" of her child, according to Hobbes. Even if this is no more than a philosophical slip resulting from the seventeenth century cultural masculinization of women in situations where they had power, Hobbes explicitly assigns power to women in a state of nature, in both *Leviathan* and *The Citizen.*[28] He therefore assigns an important fundamental identity to women as autonomous and unmarried, which, in principle, augments their status as "wifemen."

One can speculate, as Hobbes must have and as Pateman does in Hobbesian terms, about how women historically chose or were forced or coerced to give up their natural freedom as lord mothers and become subject to the dominion of their husbands, as a widespread institutional fact of society.[29] But however that account is developed, Hobbes' conception of women, be it an anthropological-historical account or a hypothetical foundation of political theory, puts women in a starting position as rights-bearing individuals, that is, as potential citizens of a state. If Hobbes' conception of women in the state of nature is accepted, then any political theory that proceeds from there is obligated either to criticize the actual condition of women when they are not full citizens or otherwise equal to men in fundamental rights, or else show why their diminished position in comparison with men is morally justified. That is, Hobbes' claim would place the burden of proof on those who wanted to continue with a status quo in which women were not "lord mothers." Hobbes himself seems to have accepted the status quo of subordinate positions for women on the basis of widespread custom.[30] In general in the seventeenth century, appeals to scripture, even by women feminists, would have been the most common form of justification for the subjugation of women in marriage.

It may be that I have just given Hobbes too much credit as a potential feminist. The honorific "lord mother" might alternatively be read as no more than a sign of the contradiction entailed in an early modern theorist imagining women in positions of power. Compared to Hobbes, Locke did not posit women as having power in a state of nature, although he did recognize their parental authority in society.[31] Locke did not have to deal with the problem raised by Hobbes in postulating equal power between men and women in a state of nature, because Locke located the institution of marriage in the state of nature, and within marriage women were already subordinate to men. Locke's explanation for long-term marriage rested on women's biological ability to conceive a second child while a first child still required care.[32] This assumed a duty of fathers to care for offspring, as well as a man's continued sexual interest in the same woman (perhaps to Locke's implicit feminist credit).

Neither Hobbes nor Locke directly addressed the foundation for the initial "conjugal right" that men had over women. In Hobbes' state of nature, it would always have required the consent of the woman. In Locke's state of nature, (hetero)sexual relations resulting in the conception of children seem simply to have been taken for granted. These relations were already assumed in Locke's comparison of human beings with other animals, which set the stage for his explanation of long-term marriage in the service of child-rearing. That is, Locke's biological approach obscures the socially constructed nature of human interactions: if human beings are animals then the functions they share with other animals need not be viewed as socially constructed any more than those of other animals. This is an interesting "sleight of concept" that is embedded in biological thinking about human beings to this day. Locke may also have assumed that in becoming a man's wife, a woman gives her consent to sexual relations with her husband once and for all (an assumption that has blocked claims of marital rape in present society). Furthermore, and perhaps most relevant historically, insofar as women were assumed to be sexually insatiable, their consent to sexual intercourse, in the state of nature or in society, could (conveniently) have been a foregone conclusion in seventeenth-century social and political theory.

Seventeenth-Century Feminism

Hilda Smith, in *Reason's Disciples,* presents convincing expository evidence that feminism, as a historical critical movement, began with a group of about fifteen seventeenth-century English women writers. Earlier work by women about women, especially in the sixteenth century, consisted of descriptions of the achievements of exceptional women in past ages. These historical *women's studies* were presented as a justification for educating exceptional upper-class women. But in the middle of the seventeenth century, Margaret Cavendish, duchess of Newcastle, Bathsua Makin, and Hannah Wooley took up Descartes' premise of the intellectual equality between the sexes to argue for education of all middle-class women, according to their vocations. By the end of the century, Mary Astell and Elizabeth Estob, who conceptualized women as a social group, asked why, given their intellectual equality, women's social situations were unequal to men's. They found the answer in the subordination of wives to their husbands and began to construct indirect arguments for greater equality in marriage.[33] Conceptually, these feminists were taking up the problem posed by Hobbes' premise of equality in a state of nature because, like Hobbes, they began from a position of theoretical equality between the sexes and sought social explanations for the actual position of women.

This type of feminist criticism as first formulated in the seventeenth century was not effective, however, until feminism was able to develop as a broad social program during the nineteenth century. Smith notes that most of the seventeenth-century feminists were royalists and Tories, as well as religiously conservative Anglicans.[34] By the end of the century, the Whigs, anti-Cartesian empiricists, and mild Anglican Latitudinarians were in control of English cultural life. While these inauspicious historical contingencies may account for the failure of feminist ideas to become popular throughout the reading public, there may have been a deeper obstacle to the success of seventeenth-century feminism, namely, the absence of positive sexual identities for women. The late seventeenth-century feminists assumed that everyone knew what they meant by the term 'woman,' and everyone still seemed to mean a "wifeman" who was colder and damper than a man, with the same sex organs as a man, only inverted. Sexual difference, insofar as it is directly relevant to reproductive function, presupposes some kind of biological foundationism. It may therefore be that the biological theories had to change, even if the change was accompanied by new false social identities for women, before women had a sufficiently distinct identity from which to launch effective resistance—and then critically turn back on that identity itself.

Nevertheless, the seventeenth century feminists did succeed in problematizing the negative social identities of "wifemen," and offering constructive alternatives based on their own experience. For instance, Mary Astell, in her Cartesian critique of custom, insisted that tradition is not itself a justification for the position of women in marriage. Although she argues by analogy, Astell's point amounts to the distinction between empirical descriptions and normative justifications: "That the Custom of the World, has put Women, generally speaking, into a State of Subjection, is not denied, but the Right can no more be prov'd from the Fact, than the Predominancy of Vice can justify it."[35]

Astell was not interested in the history of women's subjugation but in the use of reason, both as a thwarted God-given capacity in women, and as a tool for criticizing that subjugation. She argued, using Descartes' *Discourse on the Method* as a model, that women could find their own religious salvation, intellectually as well as emotionally. Her general endorsement of reason was specifically applied to the current practice of not educating women on a par with men. In *A Serious Proposal to the Ladies,* Astell argued for a college for upper-class women that would prepare them for intellectual pursuits and High Church service. She was convinced that many of the faults attributed to women were the result of their social roles and that these faults could be corrected through education. Astell intended for educated women to spend their lives in productive work, and her own experience as an unmarried woman actively engaged in intellectual work in a community of similarly inclined women in London supported her ideal.[36] The emphasis on charitable service and scholarship was a foundation for an alternative female gender identity for Astell and her peers. Still, the relationships within Astell's proposed school were to be permeated by religious and scholarly love so that their benevolence toward those outside their group would improve society. And this kind of commitment to sanctified usefulness to others restrained Astell from directly condemning the subjugation of women within marriage. Although she believed that "the whole World is a single Ladys Family," she upheld marriage as a sacred institution and addressed her critique of the position of women in marriage to ways in which men might choose their wives for reasons other than material gain and temporary sexual desire; she believed that if men chose wives more wisely, in friendship, that they would treat them with more kindness after they got the worldly goods they wanted or passions cooled.[37]

Elizabeth Elstob, the first professional scholar to compile an Ango-Saxon grammar, was an example of the kind of female intellectual Astell might have been describing. In her introduction to *An English-Saxon Homily on the Birth-day of St. Gregory,* Estob argued for the value of educating women on an analogy with the usefulness of scholarly work itself.[38]

In contrast to the well-bred argument of Astell and example of Estob, the anonymous female writer of *An Essay in Defense of the Female Sex* attacked "the Usurpation of Man; and the Tyranny of Custom (here in England, especially)" more directly than any previous feminist. She used Locke's empiricist epistemology to look for social causes of the inequality between the sexes. Rather than argue that women were as good as men, she insisted that they were better because their natural differences predisposed them to intellectual superiority. Men, therefore, had conspired to keep women subordinate by depriving them of education and imprisoning them in domestic spheres. Even so, what women did domestically was more important than anything accomplished by men.[39]

From a contemporary feminist perspective, some might find the intellectualism of Astell, Elstob, and the anonymous writer problematic because of the widespread exclusion of women from seventeenth-century intellectual work, and what is now considered to be the modern gender bias of rationality and reason. All of the seventeenth-century feminists could be accused of a naive ahistoricism in their faith that reason was the great equalizer and that it could be used, through education, to correct the unjust treatment of women. However, they succeeded in shifting the social identity of women, both theoretically and in their own experience. Astell had a specific female identity as an unmarried female intellectual, and Elstob defended her identity as a working scholar. One can only speculate about the life of the anonymous writer. Her attack on married men as the deliberate oppressors of women and her defiant claim that even domestic work was proof of greater female competence suggests that she might have been a "wifeman." This polemical move against an oppressor from the standpoint of an identity that has been assigned to one as part of the conditions of oppression has the ring of much later radicalism. Before such a defiant position could be effectively mobilized in history, intervening forms of feminism, based on ways in which women were more gently (and ineffectively) superior to men, such as their spirituality, had to be constructed and overturned.

What transcultural conclusions can be drawn about the early modern monosexual foundations of gender and the de facto exclusion of women from intellectual work? How can such conclusions avoid the anachronism of applying later constructions of female gender onto seventeenth-century culture? Why are such conclusions necessary? Trans-cultural conclusions are necessary to explain how the seventeenth-century condition and conceptualization of women led to later modern conditions and conceptualizations of women as a distinct and different sex and a sexualized gender. Although women were not assigned virtues in gender terms because they were childbearers, their childbearing function seems to have been sufficient to associate them with inferior social status in relation to men. Although some women ruled and even wrote, it was generally assumed that they did these things not as women, but because of their male characteristics; and there are no cases in which female rulers or writers were recognized to be the equals or superiors of men of the same class who were engaged in similar activities. It may be that the lack of sexual foundationism for female difference was enabling to some seventeenth-century women. Contemporary feminist criticism of later modern sexual dimorphism would seem to support that interpretation. But it may also be that the close association of social status with gender in the early modern period was more damaging to female achievement than later association of female gender with biological sexual difference was to be.

At any rate, it seems necessary now to assume some kind of significant sexual dimorphism in order to refer to or pick out the group of women, as opposed to men, in the early modern period. And it seems fairly clear that women, as a conceptualized group, lacked a positive identity, in its own right, during the seventeenth century.

Twelve

Slavery without Race

Several years ago I attended a great meeting in the interest of Hampton Institute at Carnegie Hall. . . . Among the speakers were R.C. Ogden, ex-Ambassador Choate, and Mark Twain; but the greatest interest of the audience was centered in Booker T. Washington, and not because he so much surpassed the others in eloquence, but because of what he represented with so much earnestness and faith. And it is this that all of that small but gallant band of coloured men who are publicly fighting the cause of their race have behind them. Even those who oppose them know that these men have the eternal principles of right on their side, and they will be victors even though they should go down in defeat. . . . They are men who are making history and a race.

—JAMES WELDON JOHNSON, *The Autobiography of an Ex-Colored Man*

James Weldon Johnson wrote the above at the close of *Autobiography of an Ex-Colored Man,* which was first published in 1912. Johnson's fictional autobiography is the narrative of an American man of mixed black and white race who, under the pressures of being witness to a lynching and burning of a black man in the South, decides to "raise a mustache" and allow whomever he encounters to assume he is white.[1] He is a protagonist on the fringes of American racial conflicts who experiences the extent to which American racial concepts are social constructions without the natural foundations the culture attaches to them.

Contemporary analyses of the cultural determinants of the apparent biological foundations of race are different from general constructivist arguments about the subject matter of science. According to constructivism, scientifically defined entities such as sex chromosomes and atoms are projections onto reality that result from the acceptance of scientific theories. But regardless of whether the constructivist interpretation of scientific entities is accepted, race, in its folk sense in American culture, would not qualify as a scientific entity or concept. No racial essences have ever been identified by empirical science; there are no necessary and sufficient conditions that all individuals assigned to any given race have in common; and the rule of *hypodescent* or the "one-drop rule" for American black racial designation has no empirical foundation in the sciences of biology, physiology, or genetics. The American folk ideas of race derive from eighteenth- and nineteenth-century speculative, ideological anthropology that posited hierarchies of human races, with the white race on top, superior to all, aesthetically, culturally, and intellectually. Those nonempirical constructions of race, determined by what now are considered to be racist evaluations, were put together as a rationalization of modern colonialism and chattel slavery. I have discussed these issues of racial theory at length elsewhere, and they are outside the present focus on the seventeenth century—but in a way that makes them highly relevant to the seventeenth century.[2]

Once it is understood that the division of human beings into races does not have the scientific foundations it is assumed to have, it becomes clear that race as a concept in current use is a cultural artifact of the modern colonial period. There was a time before the racial paradigm had been constructed, just as there is probably coming a time when it will dissolve or shift. The second half of the seventeenth century was the first period in modern European history when there was extensive contact between Europe and those areas that, in the second half of the twentieth century, have been conceptualized as the Third World. Europeans enslaved Africans and committed genocide against indigenous Americans in the seventeenth century, but they did not have a concept of race such as came to be taken for granted by the end of the eighteenth century.

However, it is frequently, easily, and anachronistically assumed, even within current emancipatory scholarship, that Africans and American Indians were oppressed by whites because they belonged to different races.[3] The purpose of this chapter is to interrogate seventeenth-century slavery in light of the fact that Africans had not yet been racialized. What assumptions were present and what assumptions were absent, from a general human rights points of view, that could account for the acceptance of slavery at that time? What relevance do those assumptions have for future historical periods when the modern concept of race may no longer be an organizing principle for membership within groups and relations among groups? I will begin with some of the historical facts about slavery during the seventeenth century and then consider the absence of concepts of race in Locke's philosophy. It should then be possible to sketch an account of the English motives for, and acceptance of, African slavery during the seventeenth century.

Seventeenth-Century Slavery

There is little dispute among historians about the main facts. Slavery had been widely practiced throughout the ancient world. Plato and Aristotle famously approved of slavery for non-Greeks, at least. Aristotle described a slave as a "living tool" and explicitly stated that some men were inherently slaves by character and temperament. Tacitus referred to slaves among the Germans in the first century. Throughout the medieval period, Moslems enslaved non-Moslems and bought and sold a small but steady number of Africans among themselves and to Mediterranean Europe. The Catholic Church attempted to do away with slavery in Anglo-Saxon England and, during the eleventh century, the city of Bristol received particular attention as a trading center for English slaves who were exported for sale in Ireland.

In the second half of the fifteenth century, due to the efforts of Prince Henry the Navigator, the Portuguese developed an African slave trade with the Spanish, who were beginning to rely on African slave labor in their American colonies. By 1600, the Dutch began to participate, and by the mid-seventeenth century, England, Denmark, Brandenburg (representing Germany), France, Sweden, and Scotland were all involved. At first, before the establishment of large-scale plantations in the Middle Atlantic colonies, the English were not enthusiastic slave traders and owners.[4] The first African slaves were imported into England in 1562, and Elizabeth I was concerned that they had been captured against their will.[5] Richard Jobson's refusal to deal in slaves in 1621 is extensively quoted on behalf of the English at that time, who, he said, "did not deal in such commodities, neither did we buy or sell one another, or any that had our own shape."[6]

By the time of the Restoration in the 1660s, Charles II strengthened English merchant interest in competition against the Dutch through the Navigation Acts, which restricted the transportation of goods to and from the colonies to English ships. As discussed in chapter 7, this was the beginning of mercantilism as an alliance between business and the state.

When the Company of Royal Adventurers of England Trading into Africa was chartered in 1660, the chief interest was gold, seconded by elephants (for ivory). But by 1663, when the Company charter was renewed, African slaves were mentioned.[7] The Company then contracted to sell the English colonies 3,000 African slaves yearly. Due to problems in financing as a result of war with the Dutch, and competition against its monopoly from local English slaving ports such as Bristol and Liverpool, the Company sold its charter to the Royal African Company in 1672. It is estimated that from 1680 to 1700, the Royal African Company bought and sold 140,000 slaves from the West Coast of Africa, and the local English "interlopers" exported 160,000.[8]

The owners of the Royal Adventurers Company were mainly peers of the realm and members of the royal family. By contrast, the Royal African company was made up mainly of men of commerce, although King James II was a member (until he abdicated), as were eminent men of affairs and government officials such as Lord Shaftesbury and successive Lord Mayors and aldermen of London. There was a lower tier of prominent citizens and investors, which included John Locke. The stock of the Royal African Company was oversubscribed because it was believed that the Dutch could be defeated with France as an ally, there was a known demand for slave labor in the American colonies, and the Company had a strong promise of royal support. Most investments were under £1000, and during the first twenty years, cash dividends of 7 percent were paid out annually. The shares changed hands fairly easily as the price fluctuated between £125 and £191, which was probably overvalued for a speculative venture.[9]

As noted, there was substantial interference with the monopoly, and in 1698 it was officially broken with the Act to Settle the Trade to Africa, which allowed any merchant to participate after paying a 10 percent tax on cargo. Owners of vessels from smaller English towns such as Chester, Taunton, Totnes, Westbury, Minehead, and Bridgewater openly joined in the slave trade and actively campaigned against the tax on the basis of their right to free trade. In 1712, the tax was abolished and thereafter England came to dominate the African slave trade worldwide.[10]

Africans were not the only group enslaved during the seventeenth century. Oliver Cromwell allowed royalist prisoners to be sold in Bristol for shipment to English plantations; and after his war in Ireland, Irish prisoners were remanded to three Bristol merchants for sale as slaves in the West Indies. Charles II allowed peasants in Somerset to be sold into slavery after he quelled the Monmouth Rebellion. Quakers and other dissenters and nonconformists were also candidates for this treatment and, until the mid-eighteenth century, English children and adults were at times kidnapped into slavery in London and Bristol. English indentured servants were also bought and sold, although that practice subsided when wages rose and stories were circulated of abusive treatment of indentured servants by planters in the West Indies.[11]

It is clear from the foregoing that the English slave trade was a strongly emergent and widely accepted economic enterprise during the late seventeenth century. People of the highest social status approved of and invested in it, and it coexisted with early modern rights rhetoric. If any Latitudinarian or even more radical religious dissenter attempted to do anything to stop the slave trade in its early years, or wrote abolitionist tracts, no record of such protest survives. While no one claims that the enslavement of white English people and the trade in indentured servants approached the numbers, wide-scale misery, or subsequent historical influence of the late seventeenth-century African slave trade, the fact remains that the existence of white slaves conceptually underlines the separation of slavery at that time from later concepts of race. The broader question is whether in seventeenth-century philosophy of science or political theory there were any concepts of race that posited fundamental inherited differences among cultural or geographical groups. I am going to answer that question narrowly in the next section, with reference to John Locke's philosophy, because Locke was the most influential empirical philosopher during and after the time when England was moving into her dominance of the world slave trade in Africans.

Locke and Race

In the late 1660s Locke assisted his patron, Lord Shaftesbury, by serving as secretary to the Lords Proprietors of Carolina. He was instrumental in drawing up *The Fundamental Constitution for the Government of Carolina,* which, among other political provisions, allowed for slavery in that colony. The *Fundamental Constitution* was never used, and scholars continue to debate the extent of Locke's authorship beyond his secretarial services.[12] If Locke did frame the main ideas in the *Fundamental Constitution,* it would mean that intellectually, as well as practically, Locke condoned chattel slavery. If Locke condoned chattel slavery in theory, this not only underlines a poisonous strain in Enlightenment political theory (for example, the provisions for slavery in the United States Constitution, which otherwise relied on Locke), but it taints the egalitarianism of the Enlightenment at its source, in seventeenth-century political theory.[13] The *Fundamental Constitution* was not incompatible with Locke's later political writings in the *Second Treatise of Government* and *The Letter Concerning Toleration:* It provided for a "manorial" system of government with a balance between the powers of the nobility and commoners—the "lords proprietors" were the nobility—and religious toleration for all who believed in God, but not for atheists.[14] The problem in egalitarian terms is that if Locke did frame the *Fundamental Constitution,* it means that his ideas of freedom and individual rights had an even more limited domain than appears, because they included race as well as sex, class, and religion—women, those without property, and selected categories of nonbelievers were already excluded from seventeenth-century rights theories.

Even if Locke merely transcribed the *Fundamental Constitution* as a favor to patron and colleagues, there is the practical evidence of his acceptance of slavery and his intention to profit from it. After an earlier investment in a company trading in the Bahamas that had been organized by Lord Shaftesbury, Locke invested a total of £600 in the Royal African Company in 1674 and 1675.[15] His investments took place while he was secretary to the Council of Trade and Plantations, a post he was removed from due to his political involvements against Charles II (as noted in chapter 7).

However, the biographical facts about Locke's connections with slavery do not necessarily mean that he was a racist as racism is presently understood; and if he were a racist in his personal views and practical life, it would not necessarily mean that his philosophy was what would now be understood as racist. That is, the question of racism in Locke's philosophy has to be answered through interpretations of his philosophical work. If racism is a type of belief that human beings are naturally divided into races, based on biological differences, and that the resulting racial differences are important because one race is mentally and physically superior or inferior to another, then racism presupposes a belief in the existence of races as groups or breeds of people with general, real, and definable biological differences. I don't think that Locke's philosophy allowed for a belief in the existence of races, from which it follows that his philosophy could not have been racist. The support for this lies specifically in at least three places: the nature of Locke's explicit repudiation of slavery in the *Second Treatise;* his remarks about the morphology of persons in his comments on "monsters" and in his theory of personal identity, in the *Essay;* and one of his reductio ad absurdum arguments against civil discrimination in the *Letter Concerning Toleration*. More generally, the doctrine that there are no innate ideas seems to undermine notions of racial difference. I'll discuss these points in turn.

Locke takes it to be a fundamental assumption of natural law that in the state of nature we may not harm one another and that, in making the transition from the state of nature to civil society, we ought not to be worse off in terms of what we peacefully possessed in the state of nature. This is why, throughout the *Second Treatise,* he faults tyranny, oppression by a sovereign, and nonrepresentative government, as signs of a state of war instigated by the sovereign. Under those conditions revolution by the people is justified, because if the sovereign has put himself at war with the people the people have a right to defend themselves.[16] Slavery imposed without consent is wrong because it is abhorrent to the individual who has the rights for which Locke argues throughout.[17] It is also wrong to enslave oneself because one does not, insofar as one is God's property, have the right to give up one's liberty—one cannot enslave oneself because one does not fully own one's own liberty.[18]

The divine ownership of human liberty conflicts with Locke's claims elsewhere in the *Second Treatise* (as discussed in chapter 6) that one does own one's liberty: life, liberty, and "estate" are all forms of property that belong to individuals in the state of nature. There are two resolutions to this conflict: God owns the individual's liberty indirectly, just as I own the "grass my horse has bit"; or, as J. P. Day suggests, God owns my liberty jointly with me (or I own it jointly with Him).[19] But whichever resolution is chosen, the inalienability of liberty holds.

According to Locke, the only clear exception to the prohibition of slavery is the enslavement of captives taken in a just war. Such captives may be enslaved as an alternative to forfeiting their lives, which they themselves choose:

> Indeed, having, by his own fault, forfeited his own Life, by some Act that deserves Death; he, to whom he has forfeited it, may (when he has it in his Power) delay to take it, and make use of him to his own Service, and he does no injury by it. For, whenever he finds the hardship of his Slavery out-weigh the value of his Life, 'tis in his Power, by resisting the Will of his Master, to draw on himself the Death he desires. (*Essay,* II, iv, 23).[20]

This is more a justification of stays of execution than it is of slavery itself. Locke says nothing about people having the right to go out and enslave other people without provocation for profit, nor does he say they have not that right. He assumes that any aggression across nations will be just, presumably undertaken in self-defense (or trade?); and he also makes it clear that punitive and compensatory damages that may be (justly) exacted by victors are limited and do not extend to the property of the wives and children of the defeated guilty parties.[21] Thus, in political theory, Locke has no foundation or justification for slavery as it was developed during the seventeenth century.

Epistemologically, Locke does not provide a basis for the racial divisions that were later used to justify slavery. Locke's remarks about "monsters" or individuals with what today would be considered severe birth defects occur in the context of his nominalist arguments that we know nothing about substances as the cohesive sources of the common attributes of all members of a natural kind. The only essences we know are in this sense *nominal essences,* created by the mind, existing in the mind, and imposed by the mind on reality. It is therefore arbitrary for Locke whether any given "monster" is judged to be a member of the species whose mother gave birth to it. This nominalism around the edges of natural kinds, as well as the nominalism of natural kinds themselves, precludes a belief in objective, natural differences between or among human groups.[22] Indeed, the third earl of Shaftesbury, Locke's patron's grandson, in his neo-Platonic writings on aesthetics, castigated Locke for having made it impossible to talk about species or breeds of animals (an irony since Locke had supervised his education).[23]

However, Locke's nominalism is a tricky issue in terms of race. It would be possible to accept the extreme nominalist doctrine that all of our knowledge of the world derives from decisions we make about the concepts with which to interpret experience, and still to distinguish between concepts that meet empirical scientific standards and concepts that depend on pseudoscience or fantasy, such as phrenological bumps on the head and unicorns. And on the ground of such extreme nominalism there is room to debate the empirical foundations of the modern racial paradigm.[24] But Locke never referred to anything like racial distinctions among groups of human beings in his discussions of nominal essence.

Furthermore, if Locke was not the extreme nominalist suggested above and did think that there was an objective reality to natural kind distinctions, as Michael Ayers interprets him, then his failure to mention racial distinctions among human beings underscores the absence of 'race' from his thought.[25] Lockean nominalism is further distanced from racial ontology by the use of the concept of racial essences in nineteenth century speculative anthropology. That sense of essence was never given an empirical base corresponding to the essence itself—presumably, its base was all of the individuals who were designated as having that essence—and it is not clear whether those who believed in racial essences subscribed to traditional philosophical concepts of substance, or merely used the term essence to give an illusion of intellectual profundity to racist ideologies.[26]

As discussed in chapter 5, Locke's concept of a person was limited to a subjective, religious self, accountable for eternal punishment on Judgement Day. This forensic person, described by Locke in terms of criteria for its sameness over time, did not have to be a same physical body over time. And in the midst of that discussion, by means of a third- or fourth-hand anecdote about an intelligent parrot in Brazil, Locke seems to regret the emphasis on physical appearance in definitions of humanity, which has the result that most human beings would not consider the parrot to be a person. Locke (rather slyly) presents the parrot story as an oddity that he has on the best authority and he prefaces the anecdote with a nod to custom:

> Since I think I may be confident, that whoever should see a Creature of his own Shape and Make, though it had no more reason all its Life, than a *Cat* or a *Parrot,* would call him still a *Man;* or who ever should hear a *Cat* or a *Parrot* discourse, reason, and philosophize, would call or think it nothing but a *Cat* or a *Parrot;* and say, the one was a dull irrational *Man,* and the other a very intelligent rational *Parrot.* (*Essay,* II, xxvii, 8)[27]

If bodily differences that denote differences between human beings and animals may not be enough of a foundation for personhood for Locke, it follows (a fortiori) that bodily differences within humankind could not be enough of a difference on which to determine personhood.

In the *Letter Concerning Toleration,* within a pragmatic argument to the sovereign that civil discrimination based on religious belief disrupts public peace and encourages rebellion, Locke speaks of how arbitrary grounds for discrimination politically unify those discriminated against. He shows by way of analogy how what would in later times be viewed as racial discrimination would be such an absurd injustice that it would be grounds for revolt.

> Suppose this Business of Religion were let alone, and that there were some other Distinction made between men and men, upon account of their different Complexions, Shapes, and Features, so that those who have black Hair (for example) or gray Eyes, should not enjoy the same Privileges as other Cit-

> izens; that they should not be permitted either to buy or sell, or live by their Callings, that Parents should not have the Government and Education of their own Children; that all should either be excluded from the Benefit of the Laws, or meet with partial Judges; can it be doubted but these Persons, thus distinguished from others by the Colour of their Hair and Eyes, and united together by one common Persecution, would be as dangerous to the Magistrate, as any other that had associated themselves meerly upon the account of Religion?[28]

Not only does Locke think that such discrimination against citizens based on physical appearance would be grounds for revolt, but he speaks of association as being based, not on that appearance, which would have implied a concept of race in the biological sense, but on the fact that they have been persecuted.

Locke's general arguments about the absence of innate ideas are also relevant to the question of whether he had a concept of race. It is still not clear whether Locke meant that there are no innate ideas, no innate beliefs, or no innate mental capacities.[29] But his denial of innate ideas, together with the general spirit of his refusal to countenance the inheritance of nationality or religion without some form of assent on the part of the heir, implies that there are no inherited cultural differences within human groups.[30] The position that cultural differences were nonhereditary and merely the result of historical contingencies was not explicitly formulated in mainstream scholarship until Franz Boas and Claude Levi-Strauss began in the twentieth century to criticize nineteenth-century social science concepts of race as an inherited source of culture.[31] But the foundation of that position is present in Locke's philosophy as much as it is present anywhere in the history of philosophy.

Locke's omission of a concept of race, insofar as race came to be the justificatory ideology behind colonialist and slave-owning oppression, supports a secular neo-universalist Lockean theory of human rights. Such a theory would locate the identity of persons in subjective awareness, memory, and the capacity for moral agency. All persons, everywhere, across national boundaries and regardless of their appearance or biology, would be the recipients of rights in common with all other persons. The catch in such neo-universalism, as resistance to the United Nations charter has proved, is that it would reach across national boundaries. But one has to reach across national boundaries with a theory of rights in order to be outraged by the enslavement of Africans in the seventeenth century. This brings me to the final section of this chapter, which is an attempt to answer the question of why the enslavement of Africans was acceptable to the English during the seventeenth century, even in the absence of concepts of race.

Seventeenth-Century Motives for Slavery

During the last decades of the seventeenth century, venture capital from all over England was invested in the slave trade according to the "triangular" mercantile format. Typically, a vessel with a cargo of alcohol and cheap manufactured goods, especially cloth products, would leave London, Bristol, or one of the smaller ports for the West Coast of Africa. Slaves would be loaded and transported to the Caribbean or American continental colonies, and tobacco, cotton, sugar, and other raw materials that had been planted and harvested with slave labor would be shipped back to England; finished products would be sold domestically and internationally, including to colonial markets.[32] Thus, both as as a commodity and a means for agricultural production, African slaves were an essential part of the colonialist economic system, a fact that was openly acknowledged at the time. The resulting profit also led to national "savings" in the form of positive balances of silver against Spain. In 1695, John Cary summed up the importance of the African slave trade in all these ways, as

> a Trade of the most Advantage to this kingdom of any we drive and as it were all profit, the first cost being little more than small matters of our own Manufacture for which we have in Return, Gold, Teeth [i.e., ivory], Wax and Negroes, the last whereof is much better than the first, being indeed the best Traffic that the kingdom hath as it doth occasion and give so vast an employment to our People both by Sea and Land. These are the Hands whereby our Plantations are improved, and 'tis by their Labours such great quantities of Sugar, Tobacco, Cotton, Ginger and Indigo are raised which being bulky Commodities imploy great numbers of our ships for Transporting hither, and the greater number of Handicraft Trades at home, spends more of our Products and Manufactures and makes more sailors who are maintained by a separate Employ—this trade indeed is our silver mines for by the overplus of our negroes above what will serve our plantations we draw great quantities thereof from the Spaniards, a trade we are lately fallen into by a compact of the two nations.[33]

Half of all African slaves died in various stages of their transportation and early bondage during the late seventeenth century, and if there was any concern for them it was at best crudely economic. African slaves were treated without regard for their own property in the Lockean sense of "life, liberty and estate." These Africans were, in English eyes, without property at a time in history when ownership and acquisition was the primary basis for human rights and well-being. The absence of later biological notions of race does not lessen the atrocity of seventeenth-century African slavery, and it does not mitigate any fundamental moral evaluation of the attitudes of slave traders, owners, and investors. So why should it matter if race as it later came to be conceptualized was absent when modern slavery got underway? First, it suggests that socially constructed racial difference may not be the main cause for violating the rights of groups of people who can be racially conceptualized. This general point is useful for analyzing later contexts of rights violations, when concepts of race are entrenched. Second, the absence of later concepts of race in the seventeenth century leads us to a deeper historical understanding of the causes of African slavery at that time, namely, religious narrowness, a strong desire for monetary gain, nationalism, and, of course, the material ability to impose slavery on others. In a word, *appropriation* was a fundamental component of *English,* God-given identity at that time.

Religious narrowness was strongest in Protestant countries, in general. For part of the sixteenth century, Spain hesitated in her policy of enslaving Africans, and the Vatican had a history of opposing slavery, especially if slaves were Catholics.[34] We have seen in chapter 7 that even the most tolerant of English Protestants, including the Latitudinarians and Locke, did not advocate the extension of toleration to Catholics, non-Christians, or atheists.

As slavery picked up steam in the seventeenth century, there was widespread popular curiosity about the cultural and physical differences of Africans. The writings of fourteenth-century Arab historians, such as Ibn Khaldun, who described Africans slaves as having a "low degree of humanity" and being in "close proximity to the animal stage" began to circulate. In contemporary travel literature, African men and women were described as lascivious "beasts" whose sexual organs were much larger than those of Europeans.[35] But these were not yet racial constructions as, for example, was David Hume's speculation that Africans were of a different human species than Europeans.[36] The prurient fascinations that later became part of concepts of race did not directly motivate the acceleration of slavery in the seventeenth century. No one in seventeenth-century England wrote that Africans deserved to be enslaved or that enslavement by Europeans was their proper position in life, because of the way they inherently were. Rather, it was the way Englishmen were, before God, and in relation to the inhabitants of other nations, that gave them an assumed right to enslave Africans.

In 1663, when the charter for the Company of Royal Adventurers into Africa was renewed, the slave trade was mentioned as an objective for the first time.[37] The year before, Charles II had granted a charter to the Royal Society of London for the Improvement of Natural Knowledge. Both organizations had the highest national approval, and the Adventurers were expected to fight against the Dutch to secure African slave trade for the English, which they did. By the time the Royal African Company got underway, most of its members were men of business. As we have seen, some of its membership overlapped with the membership of the Royal Society, but it was the historians of the scientific organization who have left a lasting record of the spirit of English nationalism at that time. The English saw themselves as possessed of a special genius: Joseph Glanvill thought that England, as a nation, had a guardian angel; George Saville believed that there was a "Natural Reason of State" that admitted no opposition and was "supported by nature."[38] Thomas Sprat, in his *History of the Royal Society,* not only envisioned a worldwide conquest by English sciences but spoke of a particular national intellectual spirit:

> If there can be a true character given of the *Universal Temper* of any Nation under Heaven; then certainly this must be ascrib'd to our Countrymen: that they have commonly an unaffected sincerity; that they love to deliver their minds with a sound simplicity; that they have the middle qualities, between the reserv'd subtle Southern, and the rough unhewn Northern people: that they are not extreamly prone to speak: that they are more concern'd, what others will think of the strength, than of the fineness of what they say: . . . which are all the best endowments, that can enter into *Philosophical Mind.* So that even the position of our climate, the air, the influence of the heavens, the composition of the English blood; . . . seem to joyn with the labours of the *Royal Society* to render our Country, a Land of *Experimental Knowledge.*[39]

Some of the above was undoubtedly rhetoric in the interests of securing the ongoing protection and freedom of scientific inquiry during a period of internal doctrinal strife. But the same rhetoric could have been applied to the protection and freedom of slave traders. If it had been applied in that context, it would have presented the key component, namely the top, of a system of racial hierarchy. And, indeed, as many writers have noted, the colonial masters did later identify their own racial characteristics with their status as slave owners. But modern colonialism and enslavement did not begin with those hierarchical taxonomic constructions—it began with a desire for money, a narrow view of who could appropriate and own, and a willingness to kill, injure, and imprison other human beings for money. If Africans had been the subjects of seventeenth-century rights rhetoric, that is, if they had been Establishment English Protestants, then their enslavement, in Lockean terms, would have been an unjust act of war, contrary to basic principles of natural law. Removing the later lens of race from seventeenth-century slavery does not function as an apology; rather, it exposes the criminal structure of slavery, according to the political beliefs of the time. Locke's argument of captivity resulting from a just war could not have been used to justify slavery because African slaves had done nothing to English slave traders to evoke a just war against them. The only seventeenth-century justification of African slavery would have been nationalism. The vehemence with which theories of race were constructed, as slavery became more entrenched as an institution on American plantations in the eighteenth and nineteen centuries, suggests that few would have found even nationalism a convincing excuse for the ongoing enslavement of Africans.

If there is a lesson from history here it is that racial concepts and racism are unlikely to be the root causes of the oppression, enslavement, and genocide of racialized human beings. This suggests that the elimination of race and racism, essential as it is, can do no more than expose the crimes which it has been used to obfuscate. Once racial concepts and racism have been eliminated, the monumental task of enforcing ethical behavior toward all human beings can be assessed. If one looks beyond the safety of economic, social, and national boundaries, it is doubtful that globally there has been much moral progress since the seventeenth century. And a legal structure for criminalizing and punishing human rights violations across national boundaries does not yet exist. Nonetheless, language that assumes the existence of such a structure is frequently used in critiques of racial oppression in other cultures, as well as in emancipatory argument that does not accept the customary oppression within its own culture.

Thirteen

Witches and Magi

[A]lways whilst he is Young, be sure to preserve his tender Mind from all impressions and Notions of *Spirits* and *Goblins,* or any fearful Apprehensions in the dark. This he will be in danger of from the indiscretion of Servants, whose usual Method it is to awe Children, and keep them in subjection, by telling them of *Raw-Head* and *Bloody Bones,* and such other Names, as carry with them the Ideas of some thing terrible and hurtful, which they have reason to be afraid of, when alone, especially in the dark. . . . I have had those complain to me, when Men, who had been thus used when young; that though their reason corrected the wrong Ideas they had taken in, and they were satisfied, that there was no cause to fear invisible Beings more in the Dark, than in the Light, yet that these Notions were apt still upon any occasion to start up first in their preposess'd Fancies, and not to be removed without some Pains.

—JOHN LOCKE, *Some Thoughts Concerning Education*

Through Boyle's contribution to British empiricism, even the inductive Baconian sciences were purged of any overt Hermetic associations. And yet the persistence of alchemical and mystical ideas, in Newton's as well as Boyle's scientific writings, suggests that their break with the Renaissance magi tradition was not as radical as the rhetoric of the new science suggested. Magi were men who could benefit society through knowledge about hidden powers in nature that could be applied to practical crafts. Witches were (usually) women who could do the same thing on a homelier level involving human and animal well-being. Both magi and witches had the capacity to do harm, but over the seventeenth century the potential danger of magi was eradicated concomitant with the theoretical transformation of all sciences into self-proclaimed empirical disciplines, regardless of how empirical their practice was in reality. The danger posed by witches was at first tied to their association with the Devil of Christian doctrine, and then dissolved through both general skepticism about the existence of witchcraft itself and epistemological problems in confirming the existence of witchcraft in particular cases.

In both England and France, the new empiricism, by making convictions difficult to obtain in the legal system, helped women who might otherwise have been convicted of witchcraft. At the same time, these empirical standards discredited traditional knowledge that might have been preserved by witches. By contrast, the Hermetic tradition of the magi was preserved in a limited way as it was carried over, under different descriptions, into assumptions in the new sciences. It may be that the persecution of witches furnished an example of unacceptable and disreputable knowledge claims in a way that spurred empiricism on, and that the stereotypical figure of the witch worked as a kind of "shrunken head" or negative model for how natural philosophers were to present themselves. Or, it may be that as men of privileged classes, the new scientists were never in danger of being accused of an association with "evil spirits." Galileo's harassment and Descartes' resulting concerns (as discussed in chapter 2) favor the first speculation; the situation of the English bachelors of science favors the second. Many different scholarly approaches to these topics have been developed, tested, and compared. In keeping with the ongoing focus on identity in this book, this chapter addresses three aspects of seventeenth-century witchcraft: the identity or lack of identity of witches; the relation of empiricist knowledge requirements to a decline in beliefs about witchcraft; and the legacy of the magi in the new science.

The Identity and Non-Identity of Witches

During the sixteenth and seventeenth centuries, approximately 100,000 individuals were tried for witchcraft in England, the continent, and New England, and approximately 50,000 were executed. The courts of the Inquisition were milder in dispensing punishment than Protestant authorities were during the peak of the mania.[1] Traditional historical treatment of early modern witch persecutions relegated the belief in witchcraft to unfortunate superstition and ignorance.[2] But during the past two or three decades, more detailed theories from political science and sociology have been applied to the subject—for example, witch trials conducted by government authorities have been interpreted as an appropriation of religious control by the leaders of the early modern state;[3] and accusations of witchcraft have been analyzed as motivated by the guilt of those who slighted women who, in earlier, more neighborly times, had been objects of charity.[4] Also, feminist scholars have interpreted the persecution of witches in varied psychological and social ways connected with constructions of gender—for example, women were persecuted so that men could take over traditionally female work in medicine and midwifery;[5] and the persecution of witches began as a vehicle for expressing general misogyny, as well as a loathing and fear of female sexuality.[6]

The feminist analyses are empirically persuasive because, except in Finland, Iceland, Estonia, and Russia, those persecuted as witches were almost always women. The typical witch was over fifty, poor, and widowed or never married. She presumably possessed skills or exhibited behavior that went against community norms for women. As a representative of socially devalued forms of female existence, the witch would have served as a negative role model against which positive ideals of conforming, male-attached female gender could be reinforced and constructed.

When government authorities took over the prosecution of witches and the subject became a focus of educated attention, before the belief in witchcraft waned, there was great interest in the sexual relations between witches and the Devil. Merry Wiesner notes that it was at the time commonly believed that female sexuality increased with age. Heinrick Kramer and Joseph Sprenger wrote in the 1486 *Malleus Maleficarum,* which became the standard witch-hunting manual, "All witchcraft comes from carnal lust, which is in women insatiable."[7]

The question of the causes of the seventeenth-century persecution of witches is somewhat outside the issues of identity that have been the main subject of this book. In terms of identity, women suspected and convicted of witchcraft were far outside of the domain of the seventeenth-century bachelors of science, both as objects of scientific study and as members of their social and intellectual groups. Except for Francis Bacon in the context of natural histories, and Joseph Glanvill in a neo-Platonist attempt to defend religion against atheism, English empiricists ignored the subject.[8] The quote from Locke's *Education* at the beginning of this chapter suggests that witchcraft might be not only a matter he would want to dismiss himself, but something that should be kept well away from children's attention.

While government officials entertained the idea of a worldwide conspiracy of witches under the auspices of the Devil, they assigned little individual agency to the typical witch herself. The concept of witchcraft shifted from acts of harm to inherent qualities of witches that were imparted to them by the Devil, hence the search for marks of intercourse with incubi and bodily sites where familiars were given suck. Even though witches were conceptualized as having rich inner lives, during endlessly tedious interrogations by prosecutors in witchcraft investigations and trials, it was the interactions with the Devil and not the independent thoughts, desires, or feelings of the witches that interested prosecutors.[9] A reexamination of the records of such interrogations would probably support a "homosocial" reading of the learned official interest in witchcraft as a triangle in which two male subjects, the interrogator and the Devil, were indirectly engaged, through the medium of the projected sexuality of the female witch. When one adds the insight that early modern female sexuality was not, despite its perceived power, clearly distinguished from male sexuality, the movements in the triangle from interrogator, through witch, to Devil, can be read as a completely masculine activity. Much of what was extracted from witches as memories of sabbaths and intercourse with the Devil was an artifact of the situation of interrogation.[10] Study of these indoctrinated, or what today would be called "false," memories would probably reveal a great deal about the sexual fantasies and desires of male "witch hunters."

The projections of diabolical agency onto women accused of witchcraft, as well as later dismissals of witchcraft on the grounds that witches were befuddled victims of melancholic delusions, deprives the early modern "witch" of any positive occult character in her own right. The interest of prosecutors in the Devil, through the witch, further deprived the witch of active participation in her own interrogation; and the lack of a positive identity for witches becomes a conceptual vacuum due to the absence of written records by the women who were thus persecuted.[11]

Legal Empiricism and Witchcraft

There is a tendency in contemporary feminist scholarship to associate the persecution of witches with the beginnings of modern science, because witches were associated with that female disorder in nature that scientific investigation and manipulation would straighten out. However, the older scholarly tradition of connecting the demise of belief in witchcraft, at least on a learned level, with the general skepticism of the modern scientific world view contributes much to a precise account of that demise. Barbara

Shapiro's account of the positive requirements for probable knowledge among mid-seventeenth-century Latitudinarians and virtuosi, as applied to English legal procedures in dealing with accusations of witchcraft, supports a reclamation of the older scholarly tradition of the "enlightening" origins of modern Western science.[12] While questions of seventeenth-century misogyny were not directly addressed in that tradition, Shapirio's account suggests that on the subject of witchcraft and the law early modern science was not directly misogynistic.

In 1542, witchcraft and conjuring with evil spirits became a crime by English law. At that time, the emphasis was on *malificium,* but in 1563, it became punishable by death to conjur with evil spirits, even in the absence of harm done. In 1607, the death penalty required proof of a compact with the Devil, and that became the requirement for witchcraft in educated belief; on a rural community level, however, accusation of witchcraft remained focused on harm done, such as illness or death in humans or animals, damage to food and crops, and male sexual impotence.[13] The 1604 law was not repealed until 1746, but executions for witchcraft ended in the early 1680s, not only in England, but throughout Europe, as the result of evidentiary difficulties.[14]

Shapiro notes that English witchcraft prosecutions depended on the support of members of Parliament, justices of the peace, and grand juries. When that class expressed its skepticism about the existence of witches in particular cases, through increasingly careful evidentiary requirements, juries, who were composed of less educated folk, could not deliver guilty findings of fact based on their religious beliefs, superstitions, or vengeful impulses towards women they did not like.[15]

Many of the seventeenth-century intellectual arguments about witchcraft were directed toward questions of legal epistemology and avoided (evaded?) the issue of whether witchcraft itself existed. As early as 1594, Reginald Scot argued in *Discoverie of Witchcraft* that European continental legal procedures did not establish proof of witchcraft based on experience, sense information, philosophical rules, or "the word of God"; and where such proof was lacking, it was the responsibility of English judges to point it out. George Gifford took up Scot's claims that natural explanations had better evidence than many claims of witchcraft, and he insisted that while likelihood of witchcraft was sufficient for indictment, proof was required for conviction. This proof was to consist of the testimony of witnesses and confession by the witch.

Although James I attacked Scot in his 1597 *Daemonologie,* his direct experience of contradiction and fraud in witchcraft cases caused him in many instances to revise his readiness to believe in the legally punishable existence of witches. Although Bacon at first supported James's wish to eradicate witchcraft because it was a form of idolatry, he came to dismiss the 'histories' of such "marvels" as "old wives tales." In *The Triall of Witchcraft,* John Cotta applied a *via media* concept of probable knowledge to medical and legal investigations of witchcraft. By 1627, Richard Bernard in his *Guide to Grand-Jurymen in Cases of Witchcraft* advised that unless witchcraft be "very cleere," grand juries ought not to send cases to trial by juries "of simple men, who proceed too often upon relations of meer presumptions, and sometimes very weak ones too, to take away men's lives." Bernard's *Guide* continued in circulation until 1686, and the continental and traditional English reliance on confession was augmented by requirements that there be additional circumstantial evidence and pretrial evaluations of the reliability of witnesses.[16]

During this time of evidentiary tightening in the legal system, Cambridge neo-Platonists were attempting to combat the metaphysical materialism of corpuscularianism and the new mechanics. They wanted to reclaim beliefs in the existence of spirits in general and thought that the existence of bad spirits in cases of witchcraft was evidence for the existence of good spirits as well. Joseph Glanvill's 1666 *Saduscismus Triumphatus: Or, a Full and Plain Evidence Concerning Witches and Apparitions* was an attempt to provide a Baconian natural history of witchcraft. That is, Glanvill, as a member of the Royal Society, wanted to provide ordinary factual evidence for the existence of witchcraft. To this end, he collected contemporary accounts of "the world of spirits" and unsuccessfully tried to get the Royal Society to endorse further inquiries officially. While other religious empiricists, such as Robert Boyle, granted that a (very) few cases of witchcraft did seem genuine, they at the same time advised critical analyses of all evidence. An outright refutation of witchcraft, based on the importance of sense evidence, appeared in John Webster's *The Displaying of Supposed Witchcraft* in 1677. Webster argued, against Glanville, that the denial of witchcraft did not entail denial of the existence of God. Webster's work was licensed by the vice-president of the Royal Society, and like John Cotta, and Robert Boyle at times, he was a physician.[17] There is, therefore, good reason to believe that he was expressing the learned scientific consensus of his day.

These examples serve to defend seventeenth-century British empiricism against accusations of a strong connection between the new science and the persecution of witches. Although the arguments against the existence of witchcraft were conducted on the abstract ground of legal epistemology, rather than attached to the defenses of particular defendants in witch trials, they nonetheless had direct impact on official prosecutions: if there were not sufficient evidence for a grand jury to return a *Billa vera,* or if an empirically minded judge summed up evidence in an unconvincing way for a trial jury, there was no conviction. However, the evidentiary ground of the dispute was as much removed from the subject of the witch herself as the theological skeptical disputes of the sixteenth century were removed from the subject of God. The growing seventeenth-century empiricist reluctance to believe in the existence of witches can therefore be read more as information about the identities of the male intellectuals in question than as revisions of the identities of the women who were likely to be accused of witchcraft. The women in question remained "melancholic and deluded" and, presumably, sexually insatiable also.

However, the removal of the Devil from the natural order is symmetrical to the original construction, through interrogation, of a homosocial triangle of learned men, lowly women, and the Devil. The learned men no longer had any contact with spirits on the level of "low magic." But this did not mean, as they themselves explicitly claimed, that they did not believe in God, who was a spirit, or that they themselves no longer had anything to do with "spirit" in nature on the level of "high magic"—whether it went by that name or was called natural philosophy. They only ended their relationship with the Devil, and as a result, the witch, as their earlier intermediary with the Devil, became unnecessary. This suggests yet another speculative hypothesis about the causes of the early modern "witchcraft mania": In the beginning of the modern period, amid religious and political conflicts, varied religious and political authorities began a conversation with the Devil, using a certain type of woman as "translator." This may have been motivated by anxiety about social change, interest in alternative sources of power, curiosity about other realms, confusion about the meaning of the new science, or an eagerness to test expanding institutional power. When political and religious disputes stabilized and more became known about the possibilities and limitations of the new science, this conversation was no longer necessary.

The Magi Legacy

The late seventeenth-century deification of Isaac Newton was more than metaphorical, though of course it fell short of literal idolatry. As already discussed in chapter 9, Newton was a realist about the presence of God behind the visible workings of the universe, and by "God" he explicitly meant a ruler over a dominion and not a purely spiritual "soul of the world." Even though Locke, in speaking for the empiricist tradition, was skeptical about the ability of natural philosophers to arrive at knowledge of the real nature of things, in the "Epistle to the Reader" of the *Essay Concerning Human Understanding,* he made an exception for Newton and referred to him as a "Master-Builder," along with Boyle, Thomas Sydenham, and Christian Huygens. Locke also wrote a review for French readers of Newton's *Principia* in 1688 that did much to introduce the work to Europe.[18] In the first edition of the *Principia,* Edmund Halley had begun the tradition of deifying Newton in his introductory "Ode to Newton," in which he referred to "the potency of heaven-born mind."

Even as the ruler of the world, God would still be a spirit. Does this mean that Newton "conjured with spirits"? Tradition implies not, because God was more than a spirit, because Newton purported to offer secular mathematical justification and empirical support for his theories, because God was good, and most important, because as the founder of modern mechanics, Newton was—to say the least—clear of witchcraft accusations. But his postulation of gravity as action at a distance and his belief in "subtle spirit" clearly locates him in the tradition of the magi. This is how he ends the "General Scholium":

> And now we might add something concerning a certain most subtle spirit which pervades and lies hid in all gross bodies; by the force and action of which spirit the particles of bodies attract one another at near distances, and cohere, if contiguous; and electric bodies operate to greater distances, as well repelling as attracting the neighboring corpuscles; and light is emitted, reflected, refracted, inflected and heats bodies; and all sensation is excited, and the members of animal bodies move at the command of the will, namely, by the vibrations of this spirit, mutually propagated along the solid filaments of the nerves, from the outward organ of sense to the brain, and from the brain into the muscles. But these are things that cannot be explained in few words, nor are we furnished with that sufficiency of experiments which is required to an accurate determination and demonstration of the laws by which this electric and elastic spirit operates.[19]

Contemporary Newtonian scholar John Henry traces Newton's concept of attractive and repulsive forces in nature to the writings of many of his older English colleagues who worked within the emerging materialistic mechanical tradition. Before the *Principia* appeared, Walter Warner referred to a "vertue radiative" as an active principle in his mechanics; Walter Charleton accepted Gassendi's Epicurean idea of inherent motion in matter; Mathew Hale wrote of a hierarchy of "active qualities" in nature. Newton himself, in a 1705 draft "Query" for the *Opticks* wrote, "We cannot say that all nature is not alive." Newton also referred to traditional medical concepts of a "vital flame" and fermentation in order to account for vital processes. Similar ideas of fermentation as an unexplained natural process that could not be accounted for in terms of inert matter were discussed by Thomas Willis, Richard Lower, and John Mayo.

Boyle was also a vitalist in this sense. Decades before the *History of the Air* appeared, he wrote privately about the presence of "aerious, ethereal, luminous spirits in all mixed bodies." As Henry notes, Boyle's spirits were both material and occult: "I began some years ago to set down a collection of some new or less heeded observations and experiments relating to the causes and effects of change in the air, which I referr'd to several heads, . . . the last of which was of the occult qualities of the air, supposing there be any such."[20] Robert Hooke also assumed that there were active vibrations of matter in the interplanetary aether. In his *Micrographia* he used occult concepts of congruity and incongruity that relied on a notion of vibrating matter.[21] Henry's historical point about vital principles and occult powers in scientific work collateral to Newton's is that these Hermetic ideas were present in seventeenth-century mechanical philosophy before Newton's recognized achievement of uniting Hermetic principles with the mechanical philosophy; furthermore, the use of such concepts can be read independently of any distant or indirect influence of alchemical and (Cambridge) neo-Platonic work.

Henry accounts for the presence of vitalism in English empiricism on the grounds that positing such forces forestalled charges of atheism against the mechanical philosophers, so that experimental study of vital forces became an important defense and support of empiricism generally.[22] However, it may also be the case that the belief in active and animate forces in matter had epistemic value in its own right. For example, Bacon's photochemical discussions of matter contained many allusions to sixteenth-century Hermetic concepts of sexual reproduction and nourishment by the spirits of metals, and organic connections between the spirit of the earth and more refined celestial spirits. As C. W. Lemmi argued in 1933, Bacon related his use of such concepts directly to astrology and ancient Greek mythology.[23]

In the context of witchcraft and scientific identity, these historical continuities are important for two reasons. First, they underscore ways in which "knowledge of the ancients" was incorporated into English empiricism through the Hermetic tradition, a tradition that remained in the background of early modern science at least through Newton's ideas of gravity and fermentation. Second,this continuity can be used to understand the layers of a hierarchy of the identity of knowers. At the bottom was the witch, typically a poor woman with no scientific equipment who applied her knowledge to ordinary, domestic-type problems. Next up would be a magician, typically a male artisan or craftsman who supplied recipes associated with witchcraft for a more sophisticated and wealthy clientele than the neighbors of a rural witch.[24] The next layer would be occupied by a male magus, who could devote more time to his occult studies, as an activity undertaken for no immediate practical gain, except perhaps the transmutation of base metals into gold. The layer above would be neo-Platonists and empirical philosophers interested in "spirits," such as Joseph Glanvill. A tier above them would be occupied by otherwise empirical and secular mechanical philosophers such as Boyle and Hooke. Finally, at the apex would be Newton, beyond reproach from contamination with unseen forces due to his demonstrated secular achievements and high public recognition.

Since the witch was never credited with wit or agency apart from the influence of the Devil, who was held to take advantage of her sexual insatiability, one wonders at the implied powers of the Devil. As a spirit, the Devil was depicted as capable of casting himself into the form of a female succubus who could collect sperm from sleeping male victims, and a male incubus who could deposit that sperm in human females.[25] Here was spirit influencing matter in a sexual dynamic similar to the development of matter according to alchemical theories. Whereas the alchemist's goal was to capture spirit with matter, the Devil presumably had perfected the practice of capturing matter with spirit.

One wonders why there was no attempt among hermetic inquirers to study the actions of succubi and incubi. Fear of the Devil might have curbed such studies, and fear of ecclesiastical authorities would have censored any records of them. To this day, the intellectual occult traditions that may have preserved such records remain so deep in the shadows of respected and respectable academic scholarship that one can only speculate about the identity of their practitioners. But if popular entertainment, "occult" literary genres, and various subcultures and cults are guides, Western women gained or regained agency in occult matters after the seventeenth century. There are both male and female self-proclaimed and peer-recognized witches and sorcerers throughout contemporary society.

As everyone knows, those who purport to "conjure with spirits," both good and bad, are not held legally responsible for any resulting *maleficium* that is not directly committed. These commonly known social facts are worth mention because it is customary for scholars of witchcraft in the early modern period to write as though their subject is an isolated blip in European history. It is as though any recognition of the historical continuity of traditions of witchcraft and belief in witches would itself be tainted by the seventeenth century religious, legal, and scientific "wrongness" of that tradition, as though the tradition died when witches were no longer officially acknowledged to exist and witchcraft no longer a (punishable) crime. Perhaps a more informative account of witchcraft in the early modern period could be constructed by working "vertically" backwards through contemporary practices of witchcraft, rather than attempting to understand and explain it "horizontally" against the scientific, legal, and religious aspects of the situation then.

The magi legacy in the work of the most rigorous seventeenth-century English empiricists suggests that in theory the world was not yet an inanimate machine. This tends to exonerate early modern science, although not technology, from devitalizing designs for, or against, nature. The only clear distinctions between seventeenth-century magi and witches appear to have been class and gender. The witch stood at a dangerous intersection of class, gender, and knowledge. The same kind of knowledge, intersected by male gender and privileged economic and social class, afforded an ascendant destiny for the bachelors of science.

Fourteen

The Wealth of Nature

It is more a question of a general domain: a very coherent and very well-stratified layer that comprises and contains, like so many partial objects, the notions of value, price, trade, circulation, income, interest. This domain, the ground and object of 'economy' in the Classical age, is that of *wealth*. . . . Whereas the Renaissance based the two *functions* of coinage (measure and substitution) on the double nature of its intrinsic *character* (the fact that it was precious) the seventeenth century turns the analysis upside down: it is the exchanging function that serves as a foundation for the other two characters (its ability to measure and its capacity to receive a price thus appearing as *qualities* deriving from that *function*).

—MICHEL FOUCAULT, *The Order of Things*

From the context of contemporary ecological and egalitarian concerns it at first seems easy to criticize seventeenth-century empiricist, Latitudinarian, and parliamentarian leaders of cultural change. What gave them the right to assume that nature was God's gift to them, to be exploited for monetary gain? But this criticism becomes difficult without a persuasive contemporary foundation for noninterventionist policies toward nature. The first section of this chapter returns to the money aspect of Locke's analysis of labor and appropriation, in order understand how such exploitation was justified in the seventeenth century. The second section is a discussion of the lack of an independent identity of nature in the seventeenth century. The final section is a somewhat speculative consideration of the problematic of radical ecology—speculative because the problems with the seventeenth-century attitude toward nature remain in place today. Nonhuman natural beings are still not agents in human society and there is no widespread recognition of their rights.

The Importance of Money

Locke wrote directly on economics both before and after the Glorious Revolution, and he was a commissioner on the Board of Trade.[1] As master and then warden of the Mint for decades, Newton efficiently administered all phases of production and distribution of coinage.[2] These practical engagements of the foremost philosopher and scientist of the (English) seventeenth century embodied the importance of money in that culture. Michel Foucault analyzes economic change from the sixteenth to the seventeenth century as a shift in the conceptual location of wealth from the intrinsic value of gold and silver to money and trade.[3] In Locke's writings on the revaluation of coinage, he held fast to the necessity for a close, if not fixed, connection between the nominal value of money and the amount of silver in the coinage. But Locke also understood that money itself was a commodity that could fluctuate in price, and in *Some Considerations of the Consequences of the Lowering of Interest, and Raising the Value of Money,* he argued against lowering interest rates artificially in order directly to benefit London merchants.[4] The seventeenth-century recognition that money, as bullion and coin, was itself a commodity was based on the lesson of Spain's impoverishment following vast imports of precious metals from the New World. Prices rose in Spain as only the very rich could afford to exchange precious metals for foreign manufactured goods; and too much money in the form of increased gold and silver was available to purchase a domestic productive output that had not risen significantly during Spain's colonial hegemony.[5]

In the *Second Treatise,* Locke justified the existence of private property in an argument that rested on the goodness of labor. That interpretation, discussed in chapter 6, is again relevant here: God originally gave the earth and all of its fruits to all men in common; individual men needed private ownership of food and other necessities of survival, and God commanded men to labor in the state of nature, according to the laws of nature; by mixing labor with the resources for survival, individuals came to own all of what they labored on.

God had decreed that man should labor, presumably because he himself labored and man was cast in his image. By tying labor to God's command, Locke protects his labor theory of ownership from criticism of the claim that men deserve to own the results of their labor. For Locke, labor itself is good because God orders it. In principle, this goodness of labor would justify both the ownership of property by those who labor little and the lack of owned property by those who labor only for wages. It would also protect the antecedant privileges of those who were learning to labor to increase their wealth, that is, to "make money." In the seventeenth century (as now), making money was often an increase in wealth for those who already had wealth. Since the increase in wealth and productivity during the early modern period benefited mostly the entrepreneurial class, who had to work at new forms of management, administration, and finance in order to make money, the dynamic form of labor that led to surplus and accumulation spoke to their situation rather than to the situation of common laborers. In America, where there was no money, there would be no incentive to labor for profit. According to Locke, 'America' represented not only the geographical place in his own time but the condition of all men in the early stages of society. Without labor for profit, the earth and its fruits would remain improved only to a subsistence level, and Locke considered that to be a deplorable condition. For example, he asserted that a "king" among the American Indians was worse off than an English wage earner.[6] Locke was not alone in this opinion. In 1641, Lewis Robert had written in *Treasure of Trafficke, or a Discourse of Foreign Trade* that although the earth is the source of "all the riches and abundance of the world . . . yet it is observable, and found true by daily experience in many countries, that the true search and inquisition thereof, in these our days, is by many too much neglected and omitted."[7] And in 1691, the naturalist John Ray wrote that God had provided enough gold and silver to make money for trade. Ray described money as "an admirable contrivance for rewarding and encouraging industry or carrying on trade and commerce," and he argued that in the absence of precious metals and money people remained "brutish and savage," as were the Indians in North America.[8]

These seventeenth-century trade advocates appeared to have no idea that the earth and natural environments could be a source of human survival and well-being, without trade or what Locke called labor. The important incentive to labor for Locke was money, or rather, precious metals. Locke accepted the intrinsic value of gold and silver and allowed that its accumulation in exchange for surplus goods produced by labor justified violating the state of nature proviso against waste.[9] The moral result of Locke's theory is a dynamic of labor-ownership-exchange-wealth that locates virtue in a belief that labor has moral value and money intrinsic value.

Locke thought that enclosing land from the commons for monetary gain was a morally justified activity in the same way that gathering acorns in the state of nature was. His use of the words "enclosure" and "common" in the *Second Treatise* context of a state of nature now seems politically abstract, or ahistorical, but in the seventeenth century there was a literal enclosure of lands previously owned in common by peasants on local levels, carried out by landlords for profit. Widespread European enclosure movements during the sixteenth and seventeenth centuries, especially in Germany and England, dispossessed tenant farmers from their subsistent rural livelihoods by forcing them to become wage laborers at a deeper level of poverty.[10] The accompanying damage to rural ecologies would today be considered a degradation of nature. When those dispossessed complained and at times rioted, especially radical sect members during the English Civil War period, their arguments for retaining possession of commons were similar to Locke's *Second Treatise* arguments. They claimed a natural right to the land and its fruits as a source of livelihood. For example, inhabitants who protested the destruction of common wetlands, in the drainage of the fens outside of London for systematic agricultural use, were outraged by their deprivation, and not by the destruction of wetlands as a natural habitat.[11] But on Locke's analysis, the rights of landlords to enclose would have been superior because they were based on the additional virtue of accumulating money.

C. B. McPherson and other Marxian scholars have criticized Locke for conveniently importing the economics of a market economy, complete with the use and accumulation of money, into his hypothetical state of nature. The motive behind Locke's projection is usually described as a desire to provide a philosophical justification for the rising bourgeoisie, who were Locke's patrons and colleagues.[12] Locke may well have had that motive, as both a landlord and an investor himself who was always careful with his accounts. However, he cannot be held responsible for the wide-scale ecological changes that were already underway in full force before he wrote the *Second Treatise*. Therefore, his justification of seventeenth-century assaults on natural environments would have been after the fact, though his theories were put to use in the interests of subsequent entrepreneurial generations.

Aside from its class bias, there are internal conceptual problems with Locke's analysis of labor and money, chief of which is equivocation in the concepts of labor and the intrinsic value of gold and silver. The word labor suggests hard work for the sake of survival; after justifying ownership through labor on that level, however, Locke switches to a concept of labor that presupposes the antecedant possession of capital. That is the first equivocation.

Gold and silver acquired through barter as a store of surplus had intrinsic value. But in the seventeenth century money was already recognized as an official form of precious metal that fluctuated in price. Also, money itself, as a medium for trade, had a variable price expressed through changes in interest rates (which Locke thought ought not to be restrained by the government). Nonetheless, Locke justifies the acquisition of money in the same way that he justifies the acquisition of precious objects in a state of nature. He begins by introducing the acquisition of precious objects for a man in a state of nature, who,

> if he would give his Nuts for a piece of Metal, pleased with its colour; or exchange his Sheep for Shells, or Wool for a sparkling Pebble or a Diamond, and keep those by him all his Life, he invaded not the Right of others, he might heap up as much of these durable things as he pleased; the *exceeding of the bounds of his* just *Property* not lying in the largeness of his Possession, but the perishing of any thing uselessly in it.[13]

Next, Locke moves beyond gold and silver, explaining, "And thus *came in the use of Money,* some lasting thing that Men might keep without spoiling, and that by mutual consent Men would take in exchange for the truly useful but perishable Supports of Life."[14] The final result is the use of money as it was understood in the seventeenth century:

> This partage of things, in an inequality of private possessions, men have made practicable out of the bounds of Societie, and without compact, only by putting a value on gold and silver and tacitly agreeing in the use of Money. For in Governments the Laws regulate the right of property, and the possession of land is determined by positive constitutions.[15]

If by the term money Locke meant gold and silver throughout these pasages, he would have written, "only by putting a value on gold and silver and tacitly agreeing in the use of them." Instead, he finishes with "the use of Money," and goes on to describe conditions under governments in which prices were expressed in money rather than fine metals. He thus equivocates on the meaning of 'money,' first using it to refer to those precious objects in the state of nature, and then using it to refer to money as a medium for trade in seventeenth-century England. That equivocation again enabled Locke to shift the sanctified aura surrounding the humble virtue of labor in the state of nature to the labor of mercantile entrepreneurs and financiers. Thus the materials labored upon shifted from natural objects to gold and silver, to wage labor, to money.

The Identity of Nature in the Seventeenth Century

Seventeenth-century political theorists were generally partial to a historical or hypothetical state of nature as a starting point from which to construct normative theories of government and society. However, they rarely considered the actual state of nature in their own time. The concept of the state of nature used by Hobbes, as well as Locke, was a condition of human existence, rather than a system of land, plants and animals that existed on its own, apart from human concerns or intervention. Even Tommaso Campanella's *City of the Sun* and Johann Valeatin Andreas's *Christianapolis,* which depicted utopias in which man lived harmoniously with nature according to practices of the new science, did not conceptualize nature as a self-contained world.[16] Earlier conceptions of benign nature in the Greek Arcadian tradition and medieval pastoral personifications of nature and the earth as a woman also depicted nature as pleasant and nurturing in the service of man.[17]

There is a strong assumption in contemporary ecological criticism and in traditional critiques of industrialization that if it were realized that nature or the earth itself were alive, or that animals had conscious mental and emotional processes, then Westerners would be less destructive to nature than they have been since the early modern period. Descartes is blamed for his conception of animals as mere machines, that is, machines similar to human machines, but without souls.[18] Bacon is blamed for his proposals to interrogate and subdue nature.[19] The empiricist architects of the mechanical corpuscular theory are held responsible for conceptualizing matter as inert and "dead," in contrast to Hermetic, neo-Platonic, and Quaker vitalist theories that found their most rigorous philosophical expression in Leibniz' monadology.[20] Explanations that assign responsibility for the abuse of nature to intellectual error or illusion overlook the cruelty of human beings to other human beings, who are recognized to be alive and sentient by their oppressors. Since few ecologists hold the treatment of nature to a higher moral standard than the treatment of human beings, the ecological criticism that is directed to false or inappropriate conceptions of nature is rendered beside the point. A completely vitalistic planet, inhabited exclusively by sentient organisms, could be recognized to exist, as such, by human inhabitants who abused the planet and other sentient beings. The abusers might value only their own lives and well-being, or value all life and its well-being but value money more. More moderately, the human beings who valued both money and nonhuman sentient life could practice a variable disregard for nature, based on utilitarian calculations that always maximized their own well-being within circumstances involving nature. Indeed, that was the policy of mainstream seventeenth-century ecology.

The goal of seventeenth-century ecological concern was to preserve the well-being and wealth of those who had already benefited materially from the short-term destruction of natural resources by *conserving* those resources. Thus, proposals for change in the management of natural resources were based on purely instrumental values. For instance, the British Navy's request that the Royal Society investigate increasing shortages of wood for shipbuilding resulted in John Evelyn's *Silva*. Evelyn proposed a policy of forest conservation that included restrictions on cutting trees near waterways, protecting young trees, replanting harvested trees with seedlings, and moving the iron mills in England to New England, where there was a "surfeit of woods." Evelyn also addressed the damage resulting from untreated mining, in his *Terra,* and the problems of air pollution in London, in *Fumifugium,* with the same type of moderate conservationism.[21] Evelyn's proposals were not assiduously followed, perhaps because those who were in a position to implement them did not suffer direct financial loss from continued depletion of and damage to natural resources.

Knowledge about conservation and agricultural technique was widely accessible during the early modern period, and those who were in a position to benefit financially from this information did assume caretaker roles in relation to nature. The French government developed complex techniques for harvesting wood in cycles of the 120 years it took to grow an oak that could be used for shipbuilding.[22] The Dutch agrarian economy was made up of small independent landowners who had never been dependent on commons. They reclaimed land from the sea and used manure intensively to their own profit.[23] Seventeenth century English entrepreneurial farmers were avid readers of instructional books for improving yields of farmland, orchards and fish ponds, as well as new techniques for animal husbandry, beekeeping, plowing, harvesting, and the like.[24]

The Ongoing Problem with Nature and Radical Ecology

It is possible to construct a Bill of Rights for Children that assigns them entitlements from adults apart from how adults might benefit from their influence over children. One can imagine a correction to the lack of identity for women in the seventeenth century, by rejecting or reconstructing the subsequent identities invented for women in the modern period. One can also imagine a doctrine of rights for Europeans that would extend to the non-European inhabitants of the Americas, Africa, Asia, and the Third World. However, it is difficult to conceive of nature as a system including the land, plants, animals, and all resources that has rights apart from human beings, such that those rights are binding on human beings.

I would like to assume that nature is not part of or under the authority of a deity, and leave aside questions of whether animals think, the planet is an organism, matter is alive or spiritual, and all life is valuable, if only because positive answers are so difficult to justify in secular empirical disciplines. But even with that degree of skepticism, it should be possible to describe nature in such a way that all forms of interference with it require strict justification similar to interference with human life. However, such a description based on *deep ecology* would be no more agreeable to common sense in Western society today than it was in the seventeenth century.[25] The only difference is that today it is easier to do without an idea of God as the creator of nature and the authority behind any special position of human beings within it. We can, therefore, more forcefully than in the seventeenth century, ask what gives anyone the right to exploit nature insofar as it is a question of right (or rights).

Nature itself, in any derivation of the old sense of *natura naturans* or nature that creates itself, is a difficult concept to pin down. The individual strivings and interactions in any ecosystem can be identified, but the general characteristics of an ecosystem itself remain empirically vague. The concept of nature, as an entity above and beyond its individual components, is a theoretical construction in the same way that society is, as an entity apart from its individual members. Theorists of environmental concern assume that from some broad perspective over the correct (unspecified or variable) period of time, nature will be "in balance." But many current paradigms in the life and earth sciences seem to explain systems of natural change without recourse to models that have stable conditions of equilibrium to which all components in a system return on a cyclical basis—tectonic plate theory in geology and contemporary Darwinian theories of evolution are ready examples.[26]

Popular and intellectual ecological awareness now includes an ideal of human beings living together in natural environments so that even if plants and animals are killed, the entire ecosystem of human and nonhuman entities is "in harmony." If it is projected into the future, this ideal becomes a utopia from the standpoint of those who do not presently live that way. More often, the ideal is imagined to be or have been instantiated in so-called traditional cultures. From the reconstructed standpoint of the inhabitants of past traditional cultures, plants, animals, rocks, and rivers are all sentient, and this (imagined) traditional world view is described as "animistic." Even though contemporary Westerners cannot become prereflective inhabitants of such cultures, they may place a high moral and ecological value on such *bioregional* life styles and local experiments to recreate them.[27]

However, the stability of preindustrial European peasant life on a village level, like the stability of Native American life before European colonization and the stability of presently isolated rural life in Third World cultures, was and is a fragile stability subject to disruption by interaction with those who locate their benefits in profits from ecological destabilization. We do not know whether a traditional cultural stability that does not require a continual creation of more wealth through the exploitation of nature is the most beneficial social structure for human beings, but it does seem to be more beneficial to nature. What could the theoretical justification for the value of such nonintervention against nature be?

Aesthetic arguments for not interfering with nature probably carry the least amount of conceptual baggage, but such arguments rest on the value of beauty to human beings. These aesthetic arguments for not interfering with nature would therefore derive their force from human virtue.[28] (Such arguments may also privilege those groups of people who do not need to derive their livelihoods directly from destructive interventions with nature, because their livelihoods have already been indirectly secured by the exploitation of natural environments.) But modern Western society has not developed historically as a result of aesthetic considerations. It has developed according to changes in tools and techniques of interacting with nature that have resulted in more wealth for those who could afford to initiate the innovative practices and carry them through. Anxieties about dwindling natural resources and the probable inability of the planet to sustain human life at comparable rates of continued exploitation of nature are a small check to this development. The wealth of nature continues to be spent in order to increase the wealth of some groups of human beings, often at the expense of other groups who are less insulated from the damage to the environment.[29]

The modern (and postmodern) situation has developed along the same lines laid down by Locke: entrepreneurial labor creates ownership, which makes exchange possible for the accumulation of more money. Those who labor on a subsistence level do not accumulate wealth and probably have to work harder than their counterparts who labored for subsistence in the "hunter-gatherer" conditions of human history.[30] Nature itself, as a unifying force behind changing ecosystems, is as mysterious as Locke thought substance was, and no one has as yet succeeded in formulating an effective argument for leaving it alone. The idea that nature be left alone for its own sake is as inconceivable now as it was then. Even highly probable damage to human life is not convincing to those who need to labor for subsistence or who want to labor to make more money. Some are able to cultivate gardens, varying in size from backyards to national parks, but *natura naturans* remains unidentified as a sui generis force that those involved in production and progress must respect.

Paradoxically, it may not have been a separation from nature that problematically accelerated in the seventeenth century but a *failure to separate from nature.* This failure to separate has always been accompanied by extremes of destructive control and anxious caretaking. We are caught fast in this "codependency" because the only kind of separation effected thus far rests on an accelerating technological artificialization that uses up increasing amounts of natural resources. While the inhabitants of technologically "advanced" cultures live increasingly distant from nature, they consume and destroy disproportionate amounts of nonrenewable resources compared to inhabitants of less technologically advanced cultures, who live closer to nature. Separation from nature in a way that would decrease its destruction probably cannot be achieved through technological advance, although we do not know of any other path. We cannot go backward technologically because less technologically advanced cultural stages may be accompanied by interventions with natural environments that are even more ugly and damaging, as suggested by pollution and toxicity in "Second World" industrialization.

Again, our only nonartifical models for preserving natural environments are traditional cultures indigenous to specific geographical areas over long periods of time, in which inhabitants identify themselves as part of nature and assign what we would call personhood to natural beings. The primary task for ecological theory, now, would be to construct a personhood of such natural beings that would be persuasive to those who are presently committed to the seemingly unstoppable program of advancing technology and artificialization at the expense of natural beings. This means that nature needs an identity, a cultural construction that would effectively place limits on it as a source of human wealth. In constructing *agency* for nature at its intersection with culture, deep ecology might require the strategies associated with *radical ecology,* because existing legislation may protect the rights of nonhuman natural beings in principle only.

As with women and nonwhites in contemporary Western states, the rights of natural beings are not likely to be recognized antecedant to laws that create those rights. The descriptions of nature and natural beings in such legislation construct a theoretical rights-bearing identity for nature. But, as with race and gender, the existence of emancipatory laws for nature does not guarantee their enforcement. In recent years, radical ecological projects, issuing from deep-ecology commitments, have used media coverage and public sympathy to perpetrate "ecotage" and "monkey wrenching" with impunity in situations where environmentally protective laws have been violated. For example, in 1986–87, agents of Sea Shepherd destroyed half the Icelandic whaling fleet, which was engaged in commercial whaling against a ban by the International Whaling Commission, and no charges were filed against them. In 1990, Sea Shepherd radical ecologists destroyed the drift nets and operating equipment of Japanese fishing ships, which were killing sea birds in violation of a 1972 treaty between the United States and Japan, and no public official note was taken of the incident.[31]

The legal protection of nonhuman natural beings may therefore give them an identity as legal beings that allows for the effective agency of concerned human beings on their behalf. However, if a radical ecologist intervened to kill a human being in order to prevent the illegal loss of nonhuman natural life, it would probably be viewed very differently from intervention to save the life of a human being. Even those who place great value on nonhuman natural beings are far from assigning moral equality to all life forms. In the end, this comes down to modern Western assumptions about identity and power: We value human life above all other life forms because we and those most dear to us are human; *we* have the power to privilege our cultural, as well as our biological, lives over the lives and forms of existence of other natural beings; and we choose to exercise that power. Thus, while preservationist legislation and its official, as well as unofficial, enforcement construct some identity for nature, this identity does not have a degree of power, in the sense of agency, comparable to human identity.

Afterword

Where Do We Go from There?

There is a joke about a middle-aged woman who happens upon a frog in the woods. "Kiss me! Kiss me," says the frog, "and I'll turn into a handsome prince!"

The woman stares, entranced, but doesn't move.

"What's the matter?" asks the frog, growing impatient. "Don't you want a handsome prince?"

"I'm sorry," says the woman, "but at this point in my life I'm actually more interested in a talking frog."

—LORRIE MOORE, "Paris"

A few general conclusions about seventeenth-century philosophy and culture can be drawn from the foregoing chapters. Even given emancipatory concerns, early modern skepticism and empiricism can be read as positive programs for the advancement of their proponents, rather than primarily as oppressive programs directed against women, nonwhites, and nature. Emotional relationships between the sexes and ideas about women were radically different from later modern interactions and constructions. Materialism in the acquisitive sense was a primary moral, political, and religious value among the rising middle class. Ideas about the self were "colder" and more tied to religious punishment. Science became authoritarian when Newton carried theology into it under a different name, although there was a pre-Newtonian period of pluralism and intellectual toleration.

While such conclusions require further study, they can meanwhile be used to illuminate subjects in present cultural criticism. The reconstruction of a seventeenth-century affectively cool self can have a liberating influence on contemporary deconstructions of other-centered constructions of female gender. The intellectual toleration within pre-Newtonian seventeenth-century English science can be used to support critiques of competitive, authoritarian, and infallible models of knowledge today. The exploitation of presently racialized populations in the absence of concepts of race suggests that present oppression of these groups might be directly addressed as criminal action rather than racial oppression.

The combination of history and philosophy also suggests interesting hypotheses. Skepticism seems to be conducive to the production of empirical knowledge. The absence of modern constructions of gender and race in the seventeenth century does not mean that women and nonwhites are necessarily treated better in the absence of such constructions. Ideas about childhood, as a distinct condition, do not entail that full rights of personhood will be assigned to children.

My analyses of the careers, epistemology, and scientific progress of the seventeenth-century bachelors of science from their standpoints, rather than from the standpoints of those who have suffered from the Enlightenment legacy, has made use of the concept of identity to account for both progress and injustice. By making 'effective agency' a necessary component of 'identity,' I have used the invention of identity to explain the success of the bachelors of science, and the absence of 'identity' to explain the low cultural status and lack of power of women, nonwhites, children, and nature. This may need further explanation, and I will try to offer it here.

The term identity in this usage refers to a description of someone who is capable of action in a given domain in which action is possible, and in which there exists at least one other being who is capable of action and of recognizing the action of another. Action thus requires a dimension of intelligibility because its effectiveness depends on the recognition of an agent as able and entitled to act. In modern Western cultures, animals cannot act in this way, even if they can perform physical motions that are the same as recognized human action. This is what it means to say that animals have no social identity. In a political democracy, those who do not have the franchise have no identity, and in legal contexts, those who are insane or underage have no identity.

Within the domains of knowledge, the law, and politics in the seventeenth century, women, children, nonwhites, and natural beings did not have identities, although they were described in varied ways in those domains by those who did have identities. This is more than another way of saying that these types of individuals were oppressed or excluded because it points to what would have had to have been different for them not to have been oppressed or excluded in that context. They would have had to have been recognized members of the groups that were able to act politically, economically, and culturally, such that their actions in those groups would have been comparable to the actions of other members of those groups, from the standpoint of those other members.

This notion of identity goes beyond entrenched psychological and philosophical intuitions about identity. However, it leaves blank the historical contingencies of human existence, such as gender, race, ethnicity, and social class. But if identity can be chosen and in some cases invented, the advantage of this notion of identity is that it directly targets the sphere of goods from which those seeking empowerment are presently excluded. Empowerment can then be conceived as acquiring, using, and displaying the skills already possessed by those who have operative identities in domains of power, rather than by attempting to construct alternative domains of power.

If someone without identity is able to construct an identity for herself or for all members of a group of which she posits her self a member, she will have empowered herself within a domain. For example, a female philosopher may find herself in a group of male philosophers who do not behave toward her as though she is a philosopher. This deprives her of identity as a philosopher, in an otherwise common domain of action. The philosophical arguments and statements she makes are not accepted as philosophical, and they will not be accepted as such until her identity as a philosopher is recognized. Within the biased group, her categorization as a woman precludes her identity as a philosopher. It makes sense that women in this position might try to shift the domain of philosophers to a domain of women philosophers, but it is not evident that this strategy will enable them to achieve an initial goal of philosophical identity. A more effective strategy would be for philosophers who are women in male-biased groups simply to continue to insist on their identity as philosophers, through their practice of philosophy. This insistence may encompass feminist philosophy, but if female philosophers retreat to philosophical feminism in reaction to male bias, they may forfeit an opportunity for empowerment as philosophers—assuming that the discipline itself has value.

Similarly, when nonwhites are hired by affirmative action procedures in the academy, their immediate task may be to resist nonwhite identifications as part of their work identities because those identifications may undermine their work identities. The paradox is that philosophers who are

African Americans may be interested in developing African-American philosophy by addressing topics of race and social justice philosophically, but in order to accomplish this they may need to take up philosophical identities that are aracial (racially neutral).

If there is both a male and a white bias in philosophy, and if previous assumptions of gender and racial neutrality have been no more than a subterfuge for maintaining the privilege of these nonuniversal interests, then neutrality itself ought not to be dismissed. On what grounds can it be dismissed if it has never been practiced? A real neutrality toward gender and race might be possible for the first time only in the wake of legislation that guarantees some presence of women and nonwhites in places and fields of work. When Jean-Paul Sartre suggested that it was a "concrete liberalism" in political contexts to accept individuals from ethnic and racial minority groups—his primary example was French Jews after World War II—not as universal individuals but as members of those groups, his insight was incomplete.[1] It is more just to accept a person as a member of a previously excluded group only to the degree that there is a tendency to reject the person because she is a member of that group. From the standpoint of the new member, resistance against persistent attitudes of exclusion may require reminding other members of the group to which she has gained entry of her minority group membership. But such reminders, as correctives to covert and subtle forms of exclusion that persist after de facto and de jure exclusion have been eliminated, are no more than tactics for entry and do not constitute strategies for participation. First male Jews and then women and blacks were able to get Ph.D.s and employment in academic philosophy. Until there is full acceptance and recognition of the philosophical identities of members of recently excluded groups who are now philosophers, excluded identifications may be asserted in order to prevent denial, on the grounds of race or gender, of philosophical identity by those who already enjoy it. But once entry to the group of philosophers is secured, or whenever it is temporarily possessed, the previously excluded identifications based on race and gender ought to be irrelevant and might be obstructive for the construction of philosophical identity.

The construction of new identities for oneself, as an individual or a member of a group, requires careful consideration. The intellectual history of the seventeenth century suggests that the new identities must have possibilities for action built into them. Most of the bachelors of science discussed in this book improved their worldly positions simultaneously to constructing new identities for themselves as empirical knowers. Locke reflects this process indirectly when he restricts his discussion of the identity of persons to the sameness of human consciousness as conscience over time. For Locke, that which could have a conscience was a self-owner and, by extension, an appropriator of other things. The core of the Lockean person is responsibility, its nemesis, punishment. Newton did not have an explicit theory of human personal identity, but his identification with God as the master craftsman and ruler of the universe indirectly echoes the same themes of responsibility and ownership.

In contrast to Locke and Newton, Descartes' location of human identity in continuously thinking, nonextended substance, was both more universal and less empowered. Not everyone is in a position to perform actions that are important enough to incur divine punishment, much less to create things that have to be ruled. But almost anyone or anything can be constructed as a constantly thinking soul that need not participate in any significant cultural activity in order to "be." That is, identity in a historically dynamic sense is a more narrow concept than being. And while the form of identity that includes agency in domains of power was inaccessible to women, children, nonwhites, and nature in the seventeenth century, possibly short-lived legal safeguards have today made it attainable to women, nonwhites, and through concerned advocacy, children and nature.

The invention of new identities in ways that include agency and power is, for want of a better word, a *radical* undertaking. We tend to think of rights as freedoms and protections that can be conferred as the result of a good moral argument. But historically, the empowerment of individuals under new identities has probably always preceded their formal possession of rights. The seventeenth-century male, English, Lockean self had the power of owning itself and derivatively owning other things as part of its identity because the rising bourgeoisie was acquiring new wealth. The Protestant proprietor preceeded a Protestant rights-bearer, in political theory and reality. Analogously, the early modern empirical thinker created new forms of knowledge before a forum granting intellectual freedom was formally accorded through the early Royal Society. The historical lesson here is that the creation of identity that brings with it a new sphere of action for its bearer probably needs to be preceded by the performance of new kinds of actions, in the world, or intellectually.

Modern concepts of nonwhite race and female gender were invented in order to justify and enforce exclusions of women and racialized beings, which were already in place without justification. The construction of false biological foundations provided a way to naturalize cruelty and exploitation that was unadorned in the seventeenth century, except for the deification of its perpetrators. It is this harshness of seventeenth-century oppression, cloaked in religious pieties that secular intellectuals cannot now utter, that makes it so instructive. If we could figure out what would have motivated English slavers to desist, for example, then we would have a persuasive and motivating argument against contemporary greed and discrimination as well.

For centuries now, in modern history, naturalistic taxonomies of race and gender have intellectually and emotionally mediated (i.e., acted as "middlemen" for) the oppression of nonwhites and women. The will to oppress need not falter as the biological foundations crumble. In present so-called backlash, sexism and racism do not dissolve in the shamefulness of discrimination without justification. The lesson of the seventeenth century, also, seems to be that oppression is possible with or without biological foundations. But would-be oppressors might be thwarted by individuals and groups, of which the bachelors of science were a prime example, who have agency built into their identities.

Notes

Notes to the Introduction

1. Susan Moller Oken argues that when justice is defined as fairness, care should be taken in picking out the recipients of fairness so as not automatically to ignore women. If women have a gender identity as subordinates to men within male-headed households, they will be ignored as they have been throughout most of modern history. See Susan Moller Oken, "Justice As Fairness—For Whom?" in *Feminist Interpretations and Political Theory,* ed. Mary Lyndon Shanley and Carole Pateman (University Park: Pennsylvania State University Press, 1991), pp. 199–217.

2. Cf. Elizabeth V. Spelman, "Simone de Beauvoir and Women: 'Just Who Does She Think "We" Is?,'" in Shanley and Pateman, *Feminist Interpretations,* pp. 199–216.

Notes to Chapter One

1. René Descartes, *Meditations on First Philosophy,* in *Descartes' Philosophical Writings,* ed. Norman Kemp Smith (New York: Random House, 1958), I, pp. 176–81.

2. Ibid., II, pp. 84–86.

3. Ibid., III, pp. 87–91.

4. Henry Frankfurt, *Demons, Dreamers, and Madmen* (New York: Bobbs-Merrill, 1970), pp. 3–13, 73–74.

5. Amélie Oksenberg Rorty, "Descartes on Thinking with the Body," in *The Cambridge Companion to Descartes,* ed. John Cottingham (Cambridge: Cambridge University Press, 1992), pp. 371–92.

6. See Susan R. Bordo, *The Flight to Objectivity: Essays on Cartesianism and Culture* (Albany: State University of New York Press, 1987). Feminist psychoanalytic interpretations of seventeenth-century philosophy occur throughout this work, esp. pp. 1–33, 97–119.

7. See Carolyn Merchant, *The Death of Nature: Women, Ecology and the Scientific Revolution* (New York: Harper and Row, 1980). For extensive discussion of the vi-

olation of the mother earth model of ecology, see pp. 1–42, 99–127. For parallel interpretations of Francis Bacon's use of sexual metaphor, see Evelyn Fox Keller, *Reflections on Gender and Science* (New Haven: Yale University Press, 1985).

8. The letter to James I is from Francis Bacon, *De Dignitate et Augmentis Scientiarum,* in *Bacon Works,* ed. James Spedding, Robert Leslie Ellis, Douglas Devon Heath (London: Longmans Green, 1887), vol. 4, p. 296, cited in Merchant, *Death of Nature,* p. 168.

9. Carol Gilligan's *In a Different Voice* (Cambridge: Harvard University Press, 1982) applies the object relations work of Nancy Chodorow to feminist discussions of the construction of male gender identity. Her work has been reapplied throughout contemporary feminist scholarship.

10. See Jane Flax, "Political Philosophy and the Patriarchal Unconscious: A Psychoanalytic Perspective on Epistemology and Metaphysics," in *Discovering Reality,* Sandra Harding and Merrill B. Hintikka (Dordrecht: D. Reidel, 1983), pp. 245–82.

11. Evelyn Fox Keller, "Gender and Science," in Harding and Hintikka, *Discovering Reality,* pp. 187–205.

12. See Merry E. Wiesner, *Women and Gender in Early Modern Europe* (Cambridge: Cambridge University Press, 1993), pp. 71–72.

13. Ibid., pp. 82–114.

14. Gilligan refers to Janet Lever's studies on this point in *Different Voice,* pp. 9–10.

15. See Judith Butler, "Gender Trouble, Feminist Theory and Psychoanalytic Discourse," in *Feminism/Postmodernism,* ed. Linda Nicholson (New York: Routledge, 1990), p. 326.

16. Ibid., p. 328.

17. Ben-Ami Scharfstein, "Descartes' Dreams," *Philosophical Forum* 1, N.S., 3 (1969): 298–89.

18. For a full discussion of Descartes' dreams with references to the psychological interpretations then in vogue, see ibid., pp. 293–317.

19. Ibid., p. 315. Scharfstein is kind about Descartes (although he couldn't have appreciated such kindness, and neither would many Cartesian scholars).

20. Bordo, *Flight to Objectivity,* p. 1.

21. John Locke, *Two Treatises of Government,* ed. Peter Laslett (Cambridge: Cambridge University Press, 1991). Locke's statement of his need to make the distinction between politics and family can be found in Book II, Chap. i, and Chap. ii, Secs. 2 and 3. His argument for the distinction occurs in II, vii, 64 and 65, and viii, 116 and 117. (I return to this topic in chapter 5.)

22. Wiesner, *Women and Gender,* pp. 30–35, 82–114.

23. See Prudence Allen, R.S.M., "Soul, Body and Transcendence in Teresa of Avila," *Toronto Journal of Theology* 3, no. 2 (Fall 1987): 252–66. Allen also emphasizes Genevieve Lloyd's point, to follow, that Descartes did not intend that women

be excluded from the study and practice of his philosophy. See idem, "Descartes, the Concept of Woman and the French Revolution," in *Revolution, Violence and Equality,* ed. Yeager Hudson and Creighton Peden (Lewiston, N.Y.: Edwin Mellen Press, 1990), pp. 62–63.

24. Genevieve Lloyd, *The Man of Reason: "Male" and "Female" in Western Philosophy* (Minneapolis: University of Minnesota Press, 1984), pp. 44–48.

25. René Descartes, *Discourse on Method,* in Kemp Smith, *Descartes' Writings,* p. 93.

26. Lloyd, *Man of Reason,* pp. 44–45.

27. Frankfurt, *Demons,* pp. 118–19.

28. Wiesner, *Women and Gender,* p. 170.

29. Ibid.

30. See, e.g., the biographical information provided by Oliphant Smeaton in *The Essayes or Councels of Francis Bacon,* ed. Walter Worrall (New York: E. P. Dutton, 1900), pp. xi–lii.

31. See, e.g, the works of Alfred Dodd on the subject of the "truth" about Francis Bacon, namely: he was the real "Shakespeare"; he founded various masonic orders and wrote in codes with secret symbols; he was innocent of the bribery charge; Queen Elizabeth was his mother (as well as Essex's by a different marriage). Dodd wrote *Francis Bacon's Personal Life-Story* in 1910 (reprinted in 2 vols., London: Rider and Company, 1986), and *The Martyrdom of Francis Bacon* (London: Rider and Company, n.d.).

32. James Stevens, *Francis Bacon and the Style of Science* (Chicago: University of Chicago Press, 1975), pp. 81–82.

33. See Melissa A. Butler's discussion of Locke's feminism in this regard in "Early Liberal Roots of Feminism and John Locke on the Attack on Patriarchy," in Shaney and Pateman, *Feminist Interpretations,* pp. 74–95. On pp. 102–3, she cites Locke's letter to Elizabeth Clarke on the education of girls from John Locke, *The Correspondence of John Locke and Edward Clarke,* ed. Benjamin Rand (Cambridge: Harvard University Press, 1927).

34. See Maurice Cranston, *John Locke: A Biography* (London: Longmans Green, 1957), pp. 215–23, 449–83.

35. Wiesner, *Women and Gender,* pp. 168–69.

36. Ibid., p. 171.

Notes to Chapter Two

1. For his overview of the late sixteenth- and early seventeenth-century role of skepticism in early modern science, see Richard R. Popkin's article in *The Encyclopedia of Philosophy,* ed. Paul Edwards (New York: Macmillan, 1967), s.v. "skepticism," vol. 7, pp. 449–61. On p. 454, Popkin describes "English skepticism" as

the turn to probabilistic knowledge within the Royal Society, but he does not securely tie this turn to a parallel continental solution to pyrrhonic skepticism, namely the Gassendi-Mersenne *via media*.

2. See Jonathan Bennett, "Truth and Stability in Descartes' Treatment of Skepticism," *Canadian Journal of Philosophy,* supp. 16 (1990): 75–106.

3. During a paper he gave on this subject, titled "A God Who Can Do *Anything?*" at the University at Albany, April 1994, Bennett indicated that he was primarily concerned with what Descartes meant by 'modality' in the contemporary sense.

4. René Descartes, *Meditations on First Philosophy,* in *Descartes' Philosophical Writings,* ed. Norman Kemp Smith (New York: Random House, 1958), I, pp. 176–77.

5. Ibid., p. 181.

6. Ibid., II, pp. 182–92.

7. Descartes' reply is cited by Kemp Smith in ibid., p. 192 n. 2, as it appears in *The Philosophical Works of Descartes,* ed. and trans. Elizabeth Haldane and G.R.T. Ross (Cambridge: Cambridge University Press, 1911; reprint 1962 in two vols.), vol. 2, p. 31.

8. See R. H. Popkin, "The Skeptical Crisis and the Rise of Modern Philosophy," Parts I, II, and III, *Review of Metaphysics* 8 (1953–54): 131–51, 307–34, 499–510; and idem, "The Role of Skepticism in Modern Philosophy Reconsidered," *Journal of the History of Philosophy* 31 (1993): 501–2. Kemp Smith presents Descartes' comment from his reply to Objection II in a footnote at the end of his edition of Meditation II (see n. 7 above), thereby indicating that he did not think the philosophical topic of skepticism was uppermost in Descartes' mind. Popkin first published "The Skeptical Crisis and the Rise of Modern Philosophy" in 1953, but it wasn't until *The History of Skepticism from Erasmus to Descartes* was published in 1960 and reviewed by Charles Schmitt that Popkin's paradigm for interpreting the history of philosophy as a chronology of responses to skepticism caught hold. Although Kemp Smith downplayed Descartes' reaction to pyrrhonic skepticism in 1958, Popkin's ideas were not widely accepted at that time.

9. Richard H. Popkin, *The History of Skepticism from Erasmus to Spinoza* (Berkeley: University of California Press, 1979), pp. xiii–xx. This is the revised edition of *The History of Skepticism from Erasmus to Descartes* (New York: Van Goram, 1960).

10. Ibid., pp. 1–8, 19.

11. Ibid., pp. 39–53, 68.

12. Ibid., pp. 80–103.

13. Ibid., pp. 114–29.

14. Ibid., pp. 151–74.

15. The comparison of Popkin's use of the term 'paradigm' to Lakatos's concept of a research program is invited by Popkin's awareness that he is working within a framework for interpretation that generates further research in the history of phi-

losophy and the history of ideas. Also, Popkin dedicated *Skepticism from Erasmus to Spinoza* to the memory of Imre Lakatos. For Lakatos's brief description of a research program, see his "Science and Pseudo-Science," in *Conceptions of Science,* ed. Milton Keynes (London: Open University Press, 1974), pp. 96–102.

16. Charles B. Schmitt, "The Development of the Historiography of Skepticism from the Renaissance to Bruckner," in *Skepticism from the Renaissance to the Enlightenment,* ed. Richard H. Popkin and Charles B. Schmitt (Wiesbaden: Otto Harrossowitz, 1987), pp. 185–201.

17. Ben-Ami Scharfstein, "Descartes' Dreams," *Philosophical Forum* 1, N.S., 3 (1969): 306–7.

18. See René Descartes, "Synopsis of the Following Six Meditations," in Kemp Smith, *Descartes' Writings,* pp. 172–73; and "Meditation VI" in ibid., pp. 230–48. See also Descartes' "Passions of the Soul," part I, art. xlvii, in Haldane and Ross, *Philosophical Works of Descartes,* vol. 1, p. 353.

19. See Descartes, "Meditations I," in Kemp Smith, *Descartes' Writings,* pp. 176–77.

20. Stephen Gaukroger, *Descartes: An Intellectual Biography* (New York: Oxford University Press, 1995), pp. 336–52.

21. Kemp Smith presents Descartes as having planned the *Essais* along with the *Discourse* "Preface", the *Meditations,* and the *Principles,* as a three-stage project to win Church support. He adds that after 1644, when the negative ecclesiastical response to the *Meditations* and *Principles* was evident, Descartes was less restrained in continuing his support of scientific research—he even had a project to construct an instrument that would do for touch what the telescope had done for sight. See Kemp Smith, *Descartes' Writings,* pp. vii–xxviii.

See also Bernard Williams's biographical introduction to Descartes in *Descartes: The Project of Pure Inquiry* (New York: Penguin, 1978), pp. 15–32. Williams takes Descartes' apprehension about the Church only as far as the *Essais;* he also notes the skepticism of Descartes' correspondents' with regard to his worries (pp. 18–19).

Notes to Chapter Three

1. R. S. Woolhouse, *The Empiricists* (Oxford: Oxford University Press, 1988), pp. 71–72.

2. Concerning the neglect of Mersenne and Gassendi, see Popkin's citation of Abbé Robert Lenoble's *Mersenne ou la naissance du mechanisme,* as an example of Mersenne's importance, in Richard H. Popkin, *The History of Skepticism from Erasmus to Spinoza* (Berkeley: University of California Press, 1979), p. 278 n. 1. See also Craig B. Brush, ed., *The Selected Works of Pierre Gassendi* (New York: Johnson Reprint Corp., 1972).

3. Popkin, *History of Skepticism,* pp. 130–31.

4. Galileo, "Letter from Cardinal Bellarmine to Paolo Foscarini," in *Discoveries and Opinions of Galileo,* ed. and trans. S. Drake (New York: Doubleday Anchor, 1957), pp. 163–64.

5. For exposition and sources on Mersenne, see Popkin, *History of Skepticism,* pp. 131–37.

6. Woolhouse, *Empiricists,* p. 48; Pierre Gassendi, *Exercises in the Form of Paradoxes in Refutation of the Aristotelians,* Preface, in Brush, ed., *Selected Works,* pp. 18–30.

7. For Gassendi's claim that there was no standard for clearness and distinctness, see Gassendi, "The Fifth Set of Objections," in *The Philosophical Works of Descartes,* ed. and trans. Elizabeth Haldane and G.R.T. Ross (Cambridge: Cambridge University Press, 1911; reprint 1962 in two vols.), vol. 2, pp. 151–52.

Popkin (*History of Skepticism,* p. 291) cites the Objection of Objections as Descartes' interpretation of Gassendi's objection in "Lettre de Monsieur Des-Cartes à Monsieur C.L.R.," in *Ouvres,* ed. Charles Adam and Paul Tannery (Paris: L. Cerf, 1897–1910), pp. 211–12. For further discussion of this issue with speculation that the idea of the subjectivity of necessary truth is implicit in Descartes' text, despite his intentions, see n.2 above; and Louis Loeb, "The Priority of Reason in Descartes," *Philosophical Review* 99 (1990): 30–43.

8. For a full account of the relevant controversies and the details on Vanini, see J. S. Spink, *French Free Thought from Gassendi to Voltaire* (London: Athlone Press, 1960), pp. 3–30.

9. Gassendi, *Exercises,* Book II, Chap. V, Sec. 5, pp. 78–79.

10. Gassendi, *The Syntagma,* Part I, Book II, Chap. 5, pp. 327–35; idem, *The Rebuttals Against Descartes,* Rebuttal to Meditation II, Doubte VIII, Sec. 2, pp. 200–201 (both in Brush, *Selected Works*).

11. Francis Bacon, *The New Organon,* Part Two, "Aphorisms," Book II, ii, reprinted in Paul Edwards and Richard H. Popkin, eds., *Readings in the History of Philosophy* (New York: Collier, 1968), p. 109.

12. For a brief account of Bacon's life, interwoven with descriptions of his major works, see Maurice Cranston's article in *The Encyclopedia of Philosophy,* ed. Paul Edwards (New York: Macmillan, 1967), s.v. "Bacon," vol. 1, pp. 235–40, See also Woolhouse's chapter on Bacon in *Empiricists,* pp. 9–26. For an overview of Bacon's Great Instauration as a project, see also Mary Hesse, "Francis Bacon's Philosophy of Science," in *Essential Articles: Francis Bacon,* ed. Brian Vickers (Hamden: Archon Books, 1968), pp. 119–40.

13. See Woolhouse, *Empiricists,* pp. 16–19, on Bacon's ideas about science.

14. Cited by Cranston from *De Interpretatione Naturae Prooemium,* in *Encyclopedia of Philosophy,* s.v. "Bacon," p. 237. Dated 1603 in *Encyclopedia Britannica,* 11th ed. (Cambridge: Cambridge University Press, 1911), s.v. "Bacon," vol. 3, p. 136.

15. Francis Bacon, *Essayes or Counsels,* ed. Walter Worrall (New York: E. P. Dutton, 1900), p. 50.

16. Cranston, *Encyclopedia of Philosophy,* s.v. "Bacon," p. 236.

17. Popkin, *History of Skepticism,* p. 109.

18. Woolhouse, *Empiricists,* p. 29.

19. For an amused discussion of the incident involving burning, see "The Philosophy of Hobbes, An Essay by the Late W.G. Pogson Smith," in Thomas Hobbes, *The Leviathan,* ed. W. G. Pogson Smith (London: Oxford University Press, 1909; reprint 1958), p. xxix.

20. For a discussion of Hobbes' possible influence on Locke and Locke's motives or oversight in ignoring him in the *Two Treatises of Government,* see Peter Laslett, "Locke and Hobbes," in John Locke, *Two Treatises of Government,* ed. Peter Laslett (Cambridge: Cambridge University Press, 1991), pp. 67–91.

21. Barbara J. Shapiro, *Probability and Certainty in Seventeenth-Century England* (Princeton: Princeton University Press, 1983), pp. 83–84.

22. For Hobbes on derivation, from *De Corpore* and *Little Treatise,* see Woolhouse, *Empiricists,* pp. 29–34.

23. Ibid., p. 35.

24. See ibid., pp. 36–44, for Hobbes on indefinite causes.

25. Pogson, "Philosophy of Hobbes," in *Leviathan,* pp. vii–ix, xxv; John Aubrey, *Brief Lives,* ed. Richard Barber (Totowa, N.J.: Barnes and Noble, 1983), p. 157.

26. Throughout the *Essay,* Locke was careful not to deny the immortality of the soul. When Stillingfleet accused him of this in his theory of personal identity, where he said that the same person need not have the same body, Locke defended his theory on the grounds that the soul itself was immortal. See Locke, *Works,* vol. 2, pp. 71–79, esp. p. 73.

27. For the details of Locke's political activities and the main events in his association with Shaftesbury, see Maurice Cranston, *John Locke: A Biography* (London: Longman Green, 1952), pp. 105–29, 214–31.

28. John Locke, *An Abridgement of An Essay Concerning Human Understanding,* ed. John W. Youlton (London: J. M. Dent, 1977), Book I, pp. 1–33. (All references to and quotes from the *Essay* in this chapter are from this edition.)

29. Locke, *Two Treatises,* Book I, pp. 137–265.

30. See Locke, *Two Treatises,* Book II, Chap. vi, Secs. 52–65, pp. 303–10; *Essay,* IV, i, ii, iii, pp. 267–97, and xxi, pp. 385–87.

31. Locke, *Essay,* "Epistle to the Reader," p. xi.

32. Locke did concede that the soul was probably an immaterial substance (*Essay,* II, xxvii, 12–16, pp. 164–67), although he stressed the difficulties in conceiving of any kind of substance (*Essay,* II, xxiii, 22–24, pp. 142–43.)

33. See Cranston, *John Locke,* pp. 214–34.

34. Ibid., pp. 75–90, 264–65, 237–74.

35. Locke, *Essay,* IV, iii, 13, p. 284. By contrast, Maurice Mandelbaum argues that Locke's critical realism did not allow for the metaphysical-scientific distinction the passage seems to suggest. For example, Mandelbaum interprets Locke as working within a continuity between ordinary experience and the atomic theory.

This issue is too complex to argue here, but generally speaking, nowhere in the *Essay* does Locke suggest that any of the matters about knowledge that he is investigating can be solved by reference to empirical science. See Maurice Mandelbaum, *Philosophy, Science and Sense Perception* (Baltimore: Johns Hopkins University Press, 1964), pp. 52–56.

36. Cited from Newton's *Opticks* by Mandelbaum in *Philosophy, Science,* pp. 66–67.

37. Charlton's work is reprinted in Robert Hugh Kargon, ed., *Physiologia-Epicuro-Gassendo-Charltonia* (New York: Johnson Reprint Corp., 1964). See Editor's Introduction, pp. xiii–xxii, for discussion of Charlton's intentions and goals.

38. When the inference is not merely to what cannot be seen in fact but to what could never be seen in principle, what Mandelbaum calls the "problem of transdiction" occurs. Mandelbaum, *Philosophy, Science,* pp. 65–68.

39. Gassendi, *Syntagma,* Part II, Books I, II, in Brush, *Selected Works,* pp. 380–408.

40. Mandelbaum, *Philosophy, Science,* pp. 107–14.

41. Ibid.

Notes to Chapter Four

1. The facts about Descartes' untimely demise are usually reported in the same way; see, e.g., Bernard Williams's biographical introduction to Descartes in *Descartes: The Project of Pure Inquiry* (New York: Penguin, 1978), pp. 23–24. For Bacon, see Oliphant Smeaton, "Introduction," in Francis Bacon, *The Essays or Counsels of Francis Bacon,* ed. Walter Worrall (New York: E. P. Dutton, 1900), pp. xxv–xxvii.

2. Bacon, "Of Marriage and the Single Life," in *Essays,* pp. 28–29.

3. For Stone's general discussions of these issues, see Lawrence Stone, *The Family, Sex and Marriage in England, 1500–1800* (New York: Harper and Row, 1979), pp. 21–37, 407–22.

4. Ibid., pp. 40–43, 71–72.

5. This was the fifth largest manor house in Dorsetshire. See R.E.W. Maddison, *The Life of the Honorable Robert Boyle* (based on Thomas Birch's 1744 work) (London: Taylor and Francis, 1969), p. 60. See also John Aubrey, *Brief Lives,* ed. Richard Barber (Totowa, N.J.: Barnes and Noble, 1983), pp. 47–48.

6. See Editor's Introduction in *Descartes' Philosophical Writings,* ed. Norman Kemp Smith (New York: Random House, 1958), p. ix.

7. Maurice Cranston, *John Locke: A Biography* (London: Longmans Green, 1957), p. 8.

8. Edward Neville Da Costa Andrade, *Sir Isaac Newton* (New York: Macmillan, 1954), pp. 25–26.

9. Aubrey, *Brief Lives,* pp. 148–64.

10. Stone, *The Family,* pp. 40–43, 71–74, 408.

11. Ibid., p. 408.

12. Merry E. Wiesner, *Women and Gender in Early Modern Europe* (Cambridge: Cambridge University Press, 1993), p. 252.

13. Wiesner, *Women and Gender,* p. 226; Stone, *The Family,* pp. 310–13.

14. Aubrey, *Brief Lives,* p. 160.

15. Stone, *The Family,* p. 313.

16. Ibid. pp. 101–3.

17. Robert Darnton, *The Great Cat Massacre* (New York: Random House, 1984), pp. 145–91. Darnton describes the extensive files of the personal and social lives of the *philosophes* kept by Joseph d'Hemy, a Paris police officer and inspector of the book trade.

18. Stone, *The Family,* p. 30.

19. William Holden Hutton, *The English Church from the Accession of Charles I to the Death of Anne, 1625–1714*, reprinted from 1903 edition (New York: Macmillan, 1934) pp. 160–62.

20. Stone, *The Family,* pp. 30–35.

21. For a discussion of the distinction between science and natural philosophy in the seventeenth century, see Barbara J. Shapiro, *Probability and Certainty in Seventeenth-Century England* (Princeton: Princeton University Press, 1983), pp. 3–15, 27–28.

On the conceptual overlap between Locke's philosophy and Newton's science (which shows that at the time they were not working in distinct disciplines), see G.A.J. Rodgers, "The System of Locke and Newton," in *Essays on Early Modern Philosophy from Descartes and Hobbes to Newton and Leibniz,* ed. Vere Chappell, vol. 7, *Seventeenth-Century Natural Scientists* (New York: Garland, 1992), pp. 315–39.

22. For standard official accounts of how these honors fit into the lives in question, see *Encyclopaedia Britannica,* 11th Ed. (Cambridge: University of Cambridge Press, 1911), s.v. "Bacon," vol. 3, pp. 135–52; s.v. "Boyle," vol. 4, pp. 354–55; s.v. "Descartes," vol. 8, pp. 79–90; s.v. "Hobbes," vol. 8, pp. 445–52; s.v. "Locke," vol. 16, pp. 844–53; s.v. "Newton," vol. 19, pp. 583–52.

Hobbes' hopes from the king are mentioned by Aubrey in *Brief Lives,* pp. 153–55.

23. Bacon, "Parents and Children," in *Essays,* p. 24.

24. For this seventeenth-century meaning of 'house' see Stone, *The Family,* p. 29.

25. Ben-Ami Scharfstein, *The Philosophers, Their Lives and the Nature of Their Thought* (Oxford: Basil Blackwell, 1980), pp. 99–100.

26. Maddison, *Robert Boyle,* pp. 198–219.

27. Ibid., pp. 206–9.

28. Ibid., pp. 182–83.

29. This is my speculation. It seems plausible in the spirit of psychoanalytic bi-

ography, although I don't know of any published interpretation of Boyle's life along such lines.

30. Maddison, *Robert Boyle,* pp. 74–75, 136–37.

31. For an account of Locke's life at Oates, see Cranston, *Locke,* pp. 314–73, 449–81. Edwards's remark is cited on p. 431.

32. Scharfstein, *The Philosophers,* pp. 346–49, 354.

33. Stone, *The Family,* p. 48.

34. Scharfstein, *The Philosophers,* p. 346.

Notes to Chapter Five

1. See Anthony Storr, *Solitude* (New York: Macmillan, 1988), pp. 79–80.

2. Lawrence Stone, *The Family, Sex and Marriage in England, 1500–1800* (New York: Harper and Row, 1979), pp. 76–84.

3. Michael Ayers, *Locke, Volume II: Ontology* (New York: Routledge, 1991), pp. 278–93.

4. John Locke, *An Essay Concerning Human Understanding,* ed. Peter H. Niddich (Oxford: Oxford University Press, 1975), Book II, Chap. xxvii, Sec. 9, p. 335. (All references to and quotes from the *Essay* in this chapter are from this edition.)

5. Ibid.

6. Ibid., pp. 328–31.

7. Ibid., p. 335.

8. Ibid., p. 336.

9. Ibid., I, I, iv, 18, p. 95.

10. Ibid., p. 346.

11. See Kenneth P. Winkler, "Locke on Personal Identity," *Journal of the History of Philosophy* 29:2 (April 1991): 214–17.

12. *Essay,* p. 336.

13. Ibid., p. 337.

14. Ibid., p. 342.

15. Ibid., p. 346.

16. Edmund Law, "A Defense of Mr. Locke's Opinion Concerning Personal Identity," in John Locke, *Works* (Scientia Verlag Aalen, 1963), vol. 3, pp. 179–80.

17. Ibid., pp. 189, 201.

18. Harold W. Noonan, *Personal Identity* (London: Routledge, 1989), pp. 48–49.

19. Ayers notes that when Locke turns to the identity of living things, he adds the organizing principle of life to the equation of identity: The same living thing need not be the same mass of matter over its lifetime; the same man or human being is a different entity from the same body. (See *Essay,* II, xxvii, 4–8; and Michael Ayers, *Locke,* pp. 260–61.) However, since same persons are defined by Locke in terms of continuity of memory, it is difficult to see what their underlying changing substratum might be because he never gives any indication that he thinks there

is anything objective within persons that is analogous to the organizing principle of material particles within living bodies.

20. Perhaps the reason so many writers since Edmund Law feel constrained to attribute something "substantial" to Lockean persons is that they fail to consider seriously that 'person' for Locke may not mean what we normally assume, i.e., a secular, living human being. Although Winkler credits Law with realizing that persons are forensic for Locke, he rejects the seventeenth-century view of persons as roles, which Law drew on, and attempts to reconstruct a Lockean theory of persons as substances (albeit in a derivative sense), which could account for ordinary human identity. See Winkler, "Locke," pp. 213–16.

21. Law, "A Defense," pp. 199–201; Ayers, *Locke,* pp. 273–75.

22. Ayers, *Locke,* p. 267.

23. *Oxford English Dictionary* (Oxford: Clarendon Press, 1989), s.v. "forensic," vol. 6, p. 55.

24. *Oxford English Dictionary,* s.v. "consciousness," vol. 3, p. 756.

25. Noonan, *Personal Identity,* p. 53. For present meanings of "conscience" see *Concise Oxford English Dictionary* (Oxford: Oxford University Press, 1969), p. 258.

26. *Essay,* p. 346.

27. Ibid., p. 347.

28. Ibid., p. 344. As Noonan points out, what Locke means by "his conscience accusing or excusing him," is "his knowledge of his own acts accusing or excusing him" (*Personal Identity,* p. 53).

29. *Essay,* p. 347.

30. Ibid., p. 345.

31. Ibid., p. 341.

32. For more comprehensive discussion of Locke's meaning of 'own' and its connection with 'acknowledgement,' see Thomas Mautner, "Locke's Own," *Locke Newsletter* 22 (1991): 73–80; Naomi Zack, "Locke's Identity Meaning of Ownership," *Locke Newsletter* 23 (1992): 105–13. (This topic is more fully explored in chapter 6.)

33. See Ayers, *Locke,* pp. 266–68; Pashal Larkin, *Property in the Eighteenth Century* (London: Longmans Green, 1930), pp. 33–59; Zack, "Locke's Identity Meaning," pp. 108–9.

34. John Locke, *Two Treatises of Government,* ed. Peter Laslett (Cambridge: Cambridge University Press, 1991), Book II, Chap. vii, Sec. 85, pp. 322–23; ix, 123, p. 350; xv, 173, p. 383.

35. In saying "Yes, I did ______," the speaker would be placing himself in a position of responsibility and liability for punishment. This kind of admission in response to an accusatory question would be a performative statement that constructs the forensic self. But it is by no means clear that silent acknowledgment of past acts would be performative, a consideration for which I am grateful to Robert G. Meyers.

36. See Law, "A Defense," pp. 196–97.

37. In the beginning of his third letter to Stillingleet, Locke defends his position that the same persons for the resurrection need not have the same bodies they did in life, on the grounds that it is not clearly required that this be the case in Scripture, and that we do not know what would be entailed by such a requirement. See Locke, *Works,* vol. 2, pp. 71–79. (Locke indicates on p. 73 that the identity of souls is not a problem in terms of the Resurrection because souls are immortal and incorruptible by definition.) Stillingfleet accused Locke of omitting all ideas of substance from rational discussion and thereby undermining the doctrine of the Trinity. See also Maurice Cranston, *John Locke: A Biography* (London: Longmans Green, 1957), pp. 412–13.

Notes to Chapter Six

1. See above, chapter 5, n. 34.

2. John Locke, *Two Treatises of Government,* ed. Peter Laslett (Cambridge: Cambridge University Press, 1991), Book II, Chap. ii, Sec. 6, pp. 270–71; iv, 23–25, pp. 284–85. (All references to the Second Treatise in this chapter are from this edition.)

3. For further analysis of this implication of Locke's theory of property, see Naomi Zack, "Locke's Identity Meaning of Ownership," *Locke Newsletter* 23 (1992): 105–13.

4. Pashal Larkin, *Property in the Eighteenth Century* (London: Longmans Green, 1930), p. 52.

5. Locke, *Two Treatises,* II, ii, 6, pp. 270–71.

6. Locke, *Two Treatises,* II, v, 28, p. 289.

7. C. B. MacPherson, *The Political Theory of Possessive Individualism* (Oxford: Oxford University Press, 1962), pp. 171, 197–222; Robert Nozick, *Anarchy, State and Utopia* (New York: Basic Books, 1974), pp. 174–82.

8. Larkin, *Property,* pp. 42–43.

9. Ibid., pp. 33–38.

10. Ibid., pp. 38–49.

11. Ibid., pp. 50–51. See also F. D. Dow, *Radicalism in the English Revolution 1640–1660* (Oxford: Basil Blackwell, 1985), p. 9.

12. See Laslett's historical arguments that the *Second Treatise* was written before the Glorious Whig Revolution of 1688, in Locke, *Two Treatises,* pp. 46–66.

13. Locke, *Two Treatises,* II, v, 25, 26, pp. 285–87.

14. Ibid. See also *John Locke: Essays on the Law of Nature,* ed. W. von. Leyden (Oxford: Oxford University Press, 1958). The title of Chapter 6 is, "Are Men Bound by the Law of Nature? Yes," and Chapter 7, "Is the Binding Force of the Law of Nature Perpetual and Universal? Yes." Although this was an early work of Locke's, there is nothing in the *Second Treatise* to indicate he changed his mind on this issue.

15. Locke, *Two Treatises,* p. 285.

16. Linda Nicholson, *Gender and History* (New York: Columbia University Press, 1986), pp. 137–38.

17. Locke, *Two Treatises,* II, i, 2, 3, p. 268; vii, 64, 65, pp. 310–11; viii, 116, 117, pp. 345–46.

18. On asymmetrical parent-child duties, see Locke, *Two Treatises,* II, vii, 65–70, pp. 310–14. For Locke's pronouncement on poor children, see Larkin, *Property,* pp. 71–73 on Locke's 1697 "Report of the Board of Trade to the Lord Justices."

19. Nicholson, *Gender and History,* pp. 138–50.

20. Ibid., pp. 155–57.

21. Locke, *Two Treatises,* p. 286.

22. See Zack, "Locke's Identity Meaning," pp. 111–12.

23. Locke, *Two Treatises,* pp. 287–88.

24. Nozick, *Anarchy,* pp. 174–75.

25. These objections can be found in (for example) Alan Carter, *The Philosophical Foundations of Property Rights* (Hemel Hempstead: Harvester Wheatsheaf, 1989), pp. 13–26; Lawrence C. Becker, *Property Rights: Philosophical Foundations* (London: Routledge and Kegan Paul, 1977), p. 35; Andrew Reeve, *Property* (Houndsmills: Macmillan, 1986), pp. 128–31.

26. Locke's argument that labor justifies ownership is interpreted by MacPherson as a defense of capitalist interests in enclosure and the early factory movement. See MacPherson, *Possessive Individualism,* pp. 209–22.

27. Locke, *Two Treatises,* II, v, 31, 33, pp. 290–91.

28. Locke, *Two Treatises,* II, ii, 6–15, pp. 270–78.

29. Ibid., II, ii, 13, pp. 275–76.

30. Conal Condren, *George Lawson's Politica and the English Revolution* (Cambridge: Cambridge University Press, 1988), pp. 98–100.

31. Locke, *Two Treatises,* II, vii, 67, p. 312.

32. Ibid., II, vii, 73, 74, pp. 315–17.

33. Ibid., II, vii, 68, p. 313.

34. Ibid., II, vii, 78–81, pp. 319–21.

35. Ibid., II, vii, 82, p. 321.

36. Cited by Larkin, *Property,* p. 75.

37. Lawrence Stone, *The Family, Sex and Marriage in England, 1500–1800* (New York: Harper and Row, 1979), p. 78.

38. Francis Bacon, "Of Simulation and Dissimulation," in *The Essays or Counsels of Francis Bacon,* ed. Walter Worrall (New York: E. P. Dutton, 1900), p. 23.

39. Ibid., pp. 19–23.

40. See Bernard Williams's biographical introduction to Descartes in *Descartes: The Project of Pure Inquiry* (New York: Penguin, 1978), pp. 24–25.

41. John Aubrey, *Brief Lives,* ed. Richard Barber (Totowa, N.J.: Barnes and Noble, 1982), p. 154.

42. R.E.W. Maddison, *The Life of the Honorable Robert Boyle* (based on Thomas Birch's 1744 work) (London: Taylor and Francis, 1969), p. 171 n. 4.

43. See John Locke, "Extracts from his Common Place Book" in *Life and Letters of John Locke,* ed. Lord [Peter] King (New York: Lenox Hill, 1972), pp. 285–86.

Notes to Chapter Seven

1. John Locke, "The Epistle to the Reader," in *An Essay Concerning Human Understanding,* ed. Peter H. Niddich (Oxford: Oxford University Press, 1975), p. 7. (All references to and quotes from the *Essay* in this chapter are from this edition.)

2. Locke, *Essay,* IV, iii, 29, p. 560.

3. For example, Newton worked out his main ideas on mechanics, the infinitesimal calculus, and optics at home at Woolsthrope while Cambridge was closed due to plague in 1665–66; and the *Principia* was published through Edmund Halley's efforts in conjunction with the Royal Society. See Edward Neville Da Costa Andrade, *Sir Isaac Newton* (New York: Macmillan, 1954), pp. 48–49, 70–71.

4. Richard S. Dunn, *The Age of Religious Wars, 1559–1689* (New York: W. W. Norton, 1970), pp. 146–53.

5. Ibid., pp. 164–72.

6. Gilbert Burnet, *A History of My Own Time, with Notes by the Earls of Dartmouth and Hardwicke, Speaker Onslow and Dean Swift,* ed. M. J. Routh, 2d ed., 6 vols. (Oxford, 1833).

7. Barbara J. Shapiro, *Probability and Certainty in Seventeenth-Century England* (Princeton: Princeton University Press, 1983), p. 111.

8. Martin I. J. Griffin, Jr., *Latitudinarianism in the Seventeenth-Century Church of England,* annotated by Richard H. Popkin, ed. Lila Freedman (Leiden: E. J. Brill, 1992), pp. 3–13.

9. John H. McLachlan, *Socinianism in Seventeenth-Century England* (Oxford: Oxford University Press, 1951), pp. 1–44 and passim.

10. See ibid., pp. 320–35.

11. For descriptions of Latitudinarian beliefs see Shapiro, *Probability,* pp. 104–19, and Griffin, *Latitudinarianism.*

12. See Abraham Woodhead, *The Protestants Plea for a Socinian,* ed. Richard Kroll, The Augustan Reprint Society, Publication Number 242 (Los Angeles: University of California, 1987), pp. vi–vii.

13. William Holden Hutton, *The English Church, from the Accession of Charles I to the Death of Anne, 1625–1714,* reprinted from 1903 edition (New York: Macmillan, 1934), pp. 297–98.

14. Lord [Peter] King, *Life and Letters of John Locke* (New York: Lenox Hill, 1972), p. 196.

15. Maurice Cranston, *John Locke: A Biography* (London: Longmans Green,

1957), p. 415.

16. Woodhead, *The Protestants Plea,* Editor's Introduction, pp. viii–ix.

17. Cranston, *Locke,* p. 414.

18. Ibid., pp. 412–13.

19. Griffin, *Latitudinarianism,* p. 113.

20. John Locke, *Works* (Scientia Verlag Aalen, 1963), vol. 2, p. 74.

21. Ibid., p. 71.

22. Griffin, *Latitudinarianism,* p. 109.

23. Cranston, *Locke,* p. 416.

24. Ibid., pp. 393–95.

25. John Milton, *Areopagitica and Of Education,* ed. George H. Savine (New York: Appleton-Century-Crofts, 1951), Editor's Introduction, pp. viii–ix.

26. Milton, *Areopagitica,* pp. 17–23.

27. Ibid., pp. 25–36.

28. Ibid., Ed.'s Intro., p. viii; Cranston, *Locke,* pp. 386–87.

29. John Locke, *A Letter Concerning Toleration,* ed. James H. Tully (Indianapolis: Hackett, 1983), Editor's Introduction, pp. 4–6, 11.

30. Ibid., pp. 1–3.

31. Ibid., p. 5; Cranston, *Locke,* pp. 81–82.

32. Locke, *Letter,* Ed.'s Intro., pp. 6–7.

33. Cranston, *Locke,* p. 107.

34. Locke, *Letter,* Ed.'s Intro., pp. 10–11.

35. Ibid., William Popple, "To the Reader," p. 21.

36. Ibid., Ed.'s Intro., p. 1.

37. Ibid., pp. 23–32.

38. Ibid., pp. 40–42.

39. Ibid., p. 33.

40. Ibid., p. 38.

41. Ibid., p. 28

42. Ibid., pp. 52–53; Ed.'s Intro., pp. 10–11.

43. Ibid., Ed.'s Intro., pp. 1–2.

44. Shapiro, *Probability,* p. 114.

45. Ibid., pp. 112–17.

Notes to Chapter Eight

1. Dorothy Stimson, *Scientists and Amateurs: A History of the Royal Society* (New York: Henry Shuman, 1948).

2. Ibid., pp. 12–32; see also Edward Neville Da Costa Andrade, *A Brief History of the Royal Society* (London: The Royal Society, 1960), pp. 1–7; and, for details about the continental connections of the first fellows of the Royal Society, see

Henry Lyons, F.R.S., *The Royal Society, 1660–1940* (New York: Greenwood Press, 1968), pp. 1–9.

3. Robert K. Merton, *Science, Technology and Society in Seventeenth Century England* (New York: H. Fertig, 1970).

4. Stimson, *Scientists and Amateurs,* p. 27.

5. Ibid., pp. 39–43.

6. Ibid., pp. 47–50; on Wilkin's flexibility, see also Barbara J. Shapiro, *Probability and Certainty in Seventeenth-Century England* (Princeton: Princeton University Press, 1983), pp. 111–12.

7. Stimson, *Scientists and Amateurs,* pp. 52–55.

8. For Kuhn's thesis on the "different thinking cap" aspect of contending theories and the importance of paradigm in mature sciences, see Thomas S. Kuhn, *The Structure of Scientific Revolutions* (Chicago: University of Chicago Press, 1970), pp. 23–35, 43–52, 111–36; idem, *The Essential Tension* (Chicago: University of Chicago Press, 1977), p. xiii.

9. Kuhn, *Essential Tension,* pp. 35–41.

10. Ibid., pp. 41–49.

11. Ibid., pp. 49–52.

12. Ibid., pp. 52–58.

13. R. M. Rattansi, "The Intellectual Origins of the Royal Society," in *Essays in Early Modern Philosophy from Descartes and Hobbes to Newton and Leibniz,* vol. 7, *Seventeenth-Century Natural Scientists,* ed. Vere Chappell (New York and London: Garland, 1992), pp. 51–55.

14. Ibid., p. 59.

15. Kuhn, *Essential Tension,* pp. 54–55.

16. Robert Boyle, *The Skeptical Cymist,* ed. E. A. Moelywyn-Hughes (London: Dent, 1964), pp. 226–30.

17. A. Rupert and Marie Boas Hall, "Philosophy and Natural Philosophy: Boyle and Spinoza," in Chapell, *Essays in Early Modern Philosophy,* pp. 86–87.

18. Ibid., p. 88.

19. Ibid., p. 89.

20. Kuhn, *Essential Tension,* pp. 44–45.

21. Robert Boyle, "Certain Physiological Essays," in *The Works of Robert Boyle,* ed. Thomas Birch (London, 1772), vol. 1, p. 302; reprinted in Marie Boas Hall, *Robert Boyle on Natural Philosophy* (Bloomington: Indiana University Press, 1965), p. 124.

22. Hall, *Robert Boyle,* p. 208.

23. Shapiro, *Probability* (Princeton: Princeton University Press, 1983), p. 66.

24. Thomas Sprat, *History of the Royal Society,* ed. Jackson I. Cope and Harold Whitmore Jones (St. Louis: Washington University Press, 1958), pp. 33–34.

25. Lyons, *The Royal Society;* on Sprat's intentions, see p. vii; on royal patronage, see p. 27; on membership and funding, see pp. 76–77.

26. Ibid., pp. 41–42.

27. For a discussion of the present pluralism in philosophy, see Nicholas Rescher, "American Philosophy Today," *Review of Metaphysics* 46 (June 1993): 717–45.

28. Eve Kosofsky Sedgwick, *Between Men: English Literature and Male Homosocial Desire* (New York: Columbia University Press, 1985). Sedgwick concentrates on examples from modern literature but her general thesis is clear in the introduction.

29. Linda Jean Shepherd, *Lifting the Veil: The Feminine Face of Science* (Boston and London: Shambhala, 1993), pp. 19, 67, 71, 115.

30. Hilda L. Smith, *Reason's Disciples: Seventeenth-Century English Feminists* (Urbana: University of Illinois Press, 1982), p. 62.

31. Samuel Pepys, *Everybody's Pepys: The Diary of Samuel Pepys 1660–1669*, ed. O. F. Morshead (London: G. Bell and Sons, 1948), pp. 400–401.

Notes to Chapter Nine

1. The agreed-upon facts of Newton's life in this section are from Edward Neville Da Costa Andrade, *Sir Isaac Newton* (New York: Macmillan, 1954); and Augustus De Morgan, *Essays on the Life and Work of Newton,* ed. Philip E. B. Jourdain (Chicago and London: Open Court, 1914).

2. Herbert Butterfield, *The Origins of Modern Science* (New York: Collier Books, 1962), pp. 161–62.

3. De Morgan, *Essays,* pp. 19–20, 50–52.

4. Maurice Cranston, *John Locke: A Biography* (London: Longman Green, 1952), pp. 372–74.

5. De Morgan, *Essays,* pp. 160–71.

6. Ibid., pp. 178–79

7. Bas van Fraassen, "To Save the Phenomena," *Journal of Philosophy* 18 (Oct. 1976): 623–32.

8. Dudley Shapere, *The Encyclopedia of Philosophy,* ed. Paul Edwards (New York: Macmillan, 1967), s.v. "Newton," vol. 5, p. 490. The quotation is from Florian Cajori's edition of Andrew Motte's 1729 translation of the *Principia* (Berkeley: University of California Press, 1934), p. 549, as cited by I. Bernard Cohen in "Hypotheses in Newton's Philosophy" (see n. 14 below for full citation), p. 207, n. 4.

9. Isaac Newton, *Selections, Mathematical Principles of Natural Philosophy,* Florian Cajori revision of 1729 Andrew Motte trans. (Chicago: Henry Regnery, 1951), p. 34.

10. Thomas S. Kuhn, *The Essential Tension* (Chicago: University of Chicago Press, 1977), p. 50.

11. Ibid., pp. 189–90.

12. Paul Feyerabend, *Against Method* (London: Verso, 1988), p. 44.

13. Maurice Mandelbaum, *Philosophy, Science and Sense Perception* (Baltimore: Johns Hopkins University Press, 1964), pp. 61–63.

14. I. Bernard Cohen, "Hypotheses in Newton's Philosophy," in *Essays in Early Modern Philosophy from Descartes and Hobbes to Newton and Leibniz,* vol. 7, *Seventeenth-Century Natural Scientists,* ed. Vere Chappell (New York and London: Garland, 1992), pp. 205–26.

15. Cited in ibid., pp. 165–66.

16. Ibid., p. 167.

17. Mandelbaum, *Philosophy, Science,* pp. 73–77.

18. Newton, *Selections,* p. 44.

19. Edwin Arthur Burtt, *The Metaphysical Foundations of Modern Physical Science* (London: Routledge and Kegan Paul, 1949), pp. 286, 290–91.

20. Amos Funkenstein, *Theology and the Scientific Imagination* (Princeton: Princeton University Press, 1986), pp. 86–97; J. E. McGuire, "Existence, Actuality and Necessity: Newton on Space and Time," in Chappell, *Essays in Early Modern Philosophy,* vol. 7, 287–88.

21. Burtt, *Metaphysical Foundations,* pp. 292–93.

22. Newton, *Selections,* pp. 44–47.

23. See, e.g., Ben-Ami Scharfstein, *The Philosophers, Their Lives and the Nature of Their Thought* (Oxford: Basil Blackwell, 1980), pp. 140–41.

24. Cecil Harmsworth, *Immortals at First Hand* (London: Desmond Harmsworth, 1933), p. 177.

25. De Morgan, *Essays,* pp. 44–45.

26. Ibid., pp. 17, 62–63.

27. Ibid., pp. 41–42.

28. Ibid., p. 25.

29. For a summary of the Newton-Leibnitz controversy, see ibid., pp. 23–34, 65–102 and 187–93.

Notes to Chapter Ten

1. See editor's note to William Blake, "The Chimney Sweeper," in *English Romantic Writers,* ed. David Perkins (New York: Harcourt Brace Jovanovich, 1967), p. 54.

2. Lawrence Stone, *The Family, Sex and Marriage in England, 1500–1800* (New York: Harper and Row, 1979), pp. 254–65, 286–99.

3. James A. Axtell, ed., *The Educational Writings of John Locke* (Cambridge: Cambridge University Press, 1968), pp. 88–97.

4. The summary of the abusive conditions of childhood until the late seventeenth century is based on the historical accounts in Philippe Ariès, *Centuries of*

Childhood, trans. Robert Baldick (New York: Alfred A. Knopf, 1962), pp. 33–50, 100–133, 241–69; Stone, *The Family,* pp. 109–36, 254–86.

5. Cited in Ariès, *Centuries,* p. 260.

6. Cited in ibid, p. 101.

7. Cited in Axtell, *Educational Writings,* p. 25.

8. Ibid., pp. 36–47.

9. A full account of the pre-publication history of the Locke-Clarke correspondence is in ibid., pp. 3–18.

10. Cited in ibid., pp. 7–8.

11. From Locke's first letter to Mrs. Clarke on her daughter's education, cited in ibid., p. 8.

12. For the populist reading, see Nathan Tarcov, *Locke's Education for Liberty* (Chicago: University of Chicago Press, 1984), pp. 3–4 and passim. For the patriarchal reading, see Linda Nicholson, *Gender and History* (New York: Columbia University Press, 1986), pp. 154–58.

13. John Locke, *Some Thoughts Concerning Education,* "To Edward Clarke of Chipley, Esq.," in Axtell, *Educational Writings,* pp. 112–13.

14. Ibid., Sec. 95, p. 197.

15. Ibid., p. 51.

16. Locke's physical program is mainly in ibid., Secs. 1–33, pp. 114–38.

17. Ibid. On mothers, see Sec. 5, p. 116; S. 7, p. 119; Sec. 11, p. 123. On doctors, see Sec. 29, pp. 136–37. On servants, see Sec. 19, p. 130; Sec. 39, p. 144; Sec. 59, pp. 154–55; Sec. 64, p. 164; Sec. 107, p. 211.

18. Ibid., Sec. 177, p. 288.

19. Ibid., Sec. 83, pp. 182–83; Sec. 69, p. 165.

20. Ibid., Sec. 1, p. 114.

21. Ibid., Sec. 33, p. 138.

22. Ibid., Sec. 36, p. 140.

23. Ibid., Sec. 38, p. 143.

24. Ibid., Sec. 41–64, pp. 142–58. On the deliberate infliction of pain, see Sec. 115, pp. 224–25.

25. Ibid., Sec. 65–66, pp. 158–59.

26. Ibid., Sec. 78, pp. 177–78.

27. Ibid., Sec. 70, pp. 165–71.

28. Ibid., Sec. 93–94, pp. 191–201.

29. Ibid., Sec. 93, p. 199.

30. Ibid., Sec. 102, pp. 206–7; Sec. 108, pp. 211–12.

31. Ibid., Sec. 162–64, pp. 266–67.

32. Ibid., Sec. 177–86, pp. 288–95.

33. Ibid., Sec. 173–76, pp. 283–88.

34. Ibid., Sec. 174, p. 284

35. Ibid., Sec. 201–4, pp. 314–16.

36. Ibid., Sec. 210–11, pp. 319–22.

37. Ibid., pp. 53–70.

38. Ibid., Sec. 216, p. 325.

39. Laura M. Purdy, *In Their Best Interest?* (Ithaca and London: Cornell University Press, 1992), pp. 89, 107–15.

40. Ibid., p. 112.

41. See Stone, *Family,* p. 469; Merry E. Wiesner, *Women and Gender in Early Modern Europe* (Cambridge: Cambridge University Press, 1993), p. 43.

42. Ariès, *Centuries,* pp. 92–93, 99.

43. Purdy, *Best Interest,* pp. 115–29.

Notes to Chapter Eleven

1. Thomas Laquer, *Making Sex: Body and Gender from the Greeks to Freud* (Cambridge: Harvard University Press, 1990), pp. 7–8.

2. Queen Elizabeth said in her Tilbury Speech, "I know I have the body of a weak and feeble woman but I have the stomach of a king." See Merry E. Wiesner, *Women and Gender in Early Modern Europe* (Cambridge: Cambridge Univerity Press, 1993), p. 242. Aphra Behn wrote in the preface to her play *The Lucky Chance,* "All I ask, is the privilege for my masculine part, the poet in me." See *Selected Writings of the Ingenious Mrs. Aphra Behn,* ed. Robert Phelps (New York: Grove Press, 1950), p. 11.

3. Laquer, *Making Sex,* pp. 14–16.

4. Alice Clark, *Working Life of Women in the Seventeenth Century,* ed. Miranda Chaytor and Jane Lewis (Boston and London: Routledge and Kegan Paul, 1982) (reprinted from 1919). See especially pp. 1–14, 42–93.

5. Ibid., pp. 14–42.

6. Clark, *Working Life,* p. xxiv; Wiesner, *Women and Gender,* pp. 49–53.

7. Wiesner, *Women and Gender,* pp. 46–47, 63–64.

8. Clark, *Working Life,* p. xxv.

9. Samuel Pepys, *Everybody's Pepys: The Diary of Samuel Pepys 1660–1669*, ed. O. F. Morshead (London: G. Bell and Sons, 1948).

10. See above, chapter 4, n. 13.

11. Christina Hole, *The English Housewife in the Seventeenth Century* (London: Chatto and Windus, 1953), pp. 180–82, 188.

12. Lawrence Stone, *The Family, Sex and Marriage in England, 1500–1800* (New York: Harper and Row, 1979), pp. 169–72, 245–46, 384.

13. Wiesner, *Women and Gender,* pp. 43–44.

14. Ibid., pp. 25–27.

15. Cited in ibid., p. 46.

16. Ibid.

17. Ibid., pp. 25–27; Carolyn Merchant, *The Death of Nature: Women, Ecology and the Scientific Revolution* (New York: Harper and Row, 1980), pp. 155–62.

18. Wiesner, *Women and Gender,* pp. 44–45, 226.

19. Ibid., p. 17.

20. Ibid.; Stone, *The Family,* pp. 53–56.

21. Stone, *The Family,* pp. 395–99.

22. Ibid.

23. Phelps, *Aphra Behn,* p. 3.

24. Ibid., pp. 8–9.

25. Aphra Behn, "The Disappointment," in ibid., pp. 239–40.

26. Carole Pateman, "'God Hath Ordained to Man a Helper': Hobbes, Patriarchy and Conjugal Right," in *Feminist Interpretations and Political Theory,* ed. Mary Lyndon Shanley and Carole Pateman (University Park: Pennsylvania State University Press, 1991), pp. 53–57.

27. On the contract in which the infant, presumably by "choosing" to live, participates, see Thomas Hobbes, *The Citizen,* ed. Bernard Gert (New York: Anchor Books, 1992), Chap. IX, Secs. 2, 3, pp. 212–13.

28. See Ibid., and Pateman, "God Hath Ordained," pp. 59–62.

29. See Hobbes, *Citizen,* Chap. IX, Secs. 4, 5, 6, pp. 213–15; and Pateman, "God Hath Ordained," pp. 62–70.

30. Ibid.

31. John Locke, *Two Treatises of Government,* ed. Peter Laslett (Cambridge: Cambridge University Press, 1991), Book II, Chap. vi, Sec. 65, p. 310.

32. Ibid., Chap vii, Secs. 78–81, pp. 319–21.

33. Hilda L. Smith, *Reason's Disciples: Seventeenth-Century English Feminists* (Urbana: University of Illinois Press, 1982), pp. 3–18.

34. Ibid., pp. 10–14.

35. Cited in Smith, *Reason's Disciples,* p. 119.

36. Ibid., pp. 121–22.

37. Ibid., pp. 117–39.

38. Ibid., pp. 139–44.

39. Ibid., pp. 144–48.

Notes to Chapter Twelve

1. James Weldon Johnson, *The Autobiography of an Ex-Colored Man,* reprinted in *Three Negro Classics,* ed. John Hope Franklin (New York: Avon, 1965), pp. 497–500.

2. Naomi Zack, *Race and Mixed Race* (Philadelphia: Temple University Press, 1993); idem, "Mixed Black and White Race and Public Policy," *Hypatia* 10:1 (Feb. 1995): 120–32; idem, "Race and Philosophic Meaning," *American Philo-*

sophical Association Newsletter on Philosophy and the Black Experience 94:1 (Fall 1994): 14–20.

3. See, e.g., John Immerwahr and Michael Burke, "Race and the Modern Philosophy Course," *Teaching Philosophy* 16:1 (March 1993): 27.

4. C. M. MacInnes, *England and Slavery* (Bristol: Arrowsmith, 1934), pp. 13–21; K. G. Davies, *The Royal African Company* (London: Longmans Green, 1957), pp. 1–37.

5. W. O. Blake, *The History of Slavery and the Slave Trade* (New York: Haskell House Publishers, 1858; reprint 1969), p. 158.

6. Nigel Tattersfield, *The Forgotten Trade* (London: Jonathan Cape, 1991), p. 11.

7. MacInnes, *England and Slavery,* pp. 26–28.

8. Ibid.

9. Davis, *Royal African Company,* pp. 58–74.

10. Tattersfield, *Forgotten Trade,* p. 15.

11. MacInnes, *England and Slavery,* pp. 21–24.

12. For discussion of Locke's authorship and further sources, see J. R. Milton, "John Locke and the *Fundamental Constitution* of Carolina," *Locke Newsletter* 21 (1990): 111–34.

13. For further discussion and sources on slavery in the U.S. Constitution, see Zack, *Race and Mixed Race,* pp. 60–61.

14. Maurice Cranston, *John Locke: A Biography* (London: Longmans Green, 1966), pp. 119–20.

15. Ibid., pp. 115, 150, 155–56.

16. John Locke, *Two Treatises of Government,* ed. Peter Laslett (Cambridge: Cambridge University Press, 1991), Book II, Chap. xix, Secs. 221–43, pp. 412–28.

17. See ibid., II, ii, 2, p. 269; iv, 22, p. 283.

18. Ibid., II, iv, 23, p. 284.

19. See John Patrick Day, "Self-Ownership," *Locke Newsletter* 20 (1989): 87.

20. Locke, *Two Treatises,* II, v, 28, p. 284.

21. Locke, *Two Treatises,* II, xvi, 181–82, pp. 388–90.

22. For Locke on nominal essences insofar as they relate to natural kinds and the boundaries between natural kinds, see John Locke, *An Essay Concerning Human Understanding,* ed. Peter H. Niddich (Oxford: Oxford University Press, 1975), Book IV, Chap. iii, Sec. 5, pp. 338–40; iii, 27, pp. 414–15; iv, 27, p. 454.

23. For the reference to Locke's nominalism by Shaftesbury's grandson, see Anthony Ashley Cooper, Third Earl of Shaftesbury, *Second Characters,* reprinted in *Philosophies of Art and Beauty,* ed. Alfred Hofstadter and Richard Kuhns (Chicago: University of Chicago Press, 1976), pp. 268–69.

24. See Zack, "Race and Philosophic Meaning."

25. Ayers is careful to distinguish between Locke's nominalism and his realism. See Michael Ayers, *Locke, Volume II: Ontology* (New York: Routledge, 1991), pp. 38–42, 207–15.

26. For a while, it was erroneously thought that racial essences were transmitted through the blood. See N. P. Dubinin, "Race and Contemporary Genetics," in *Race, Science and Society,* ed. Leo Kuper (New York: Columbia University Press, 1965), pp. 71–74.

27. Locke, *Essay,* p. 333.

28. John Locke, *A Letter Concerning Toleration,* ed. James H. Tully (Indianapolis: Hackett, 1983), p. 52.

29. See Michael Ayers, *Locke, Volume I: Epistemology* (New York: Routledge, 1991), pp. 44–52, 67–70.

30. On Locke's denial that nationality is automatically inherited, see Locke, *Two Treatises,* II, vi, 58, pp. 306–7; on Locke's denial that religion is inherited see Locke, *Letter,* p. 28.

31. For details on the history of this paradigm shift in the social scenes, see Claude Levi-Strauss, "Race and History," in Kuper, *Race,* pp. 95–135; and R. Fred Wacker, *Ethnicity, Pluralism and Race* (Westport, Conn.: Greenwood, 1983).

32. MacInnes, *England and Slavery,* pp. 39–53; Davies, *Royal African Company,* pp. 16–32; Tattersfield, *Forgotten Trade,* pp. 4–8.

33. Cited in MacInnes, *England and Slavery,* p. 29, from John Cary, *An Essay towards regulating the Trade and employing the poor of the Kingdom* (London, 1717), pp. 52–53.

34. David Brion Davis, *Slavery and Human Progress* (Oxford: Oxford University Press, 1984), pp. 54–62.

35. Ibid., pp. 43–44.

36. D. P. Mannix and M. Cordley, *Black Cargoes: A History of the Atlantic Slave Trade* (New York: Viking, 1968), p. 19.

37. See Richard H. Popkin, "Hume's Racism," *Philosophical Forum* 9:2–3 (Winter–Spring 1977–78): 213.

38. Davies, *Royal African Company,* p. 41.

38. Thomas Sprat, *History of the Royal Society,* ed. Jackson I. Cope and Harold Whitmore Jones (St. Louis: Washington University Press, 1958), Editors' Introduction, p. xix.

39. Cited in ibid., p. 114.

Notes to Chapter Thirteen

1. Merry E. Wiesner, *Women and Gender in Early Modern Europe* (Cambridge: Cambridge Univerity Press, 1993), p. 219–21.

2. See, e.g., Hugh R. Trevor-Roper, *The European Witch Craze of the Sixteenth and Seventeenth Centuries and Other Essays* (Harmondsworth: Penguin Books, 1969).

3. See Christina Larner, *Enemies of God: The Witch Hunt in Scotland* (Baltimore: Johns Hopkins University Press, 1981); idem, *Witchcraft and Religion,* Oxford: Basil Blackwell, 1982.

4. See Keith Thomas, *Religion and the Decline of Magic* (London: Weidenfeld and Nicolson, 1971), pp. 552–67.

5. See Barbara Ehrenreich and Deirdre English, *Witches, Midwives and Nurses: A History of Women Healers* (New York: Feminist Press, 1973); for the now widely accepted refutation of their thesis, see David Harley, "Historians As Demonologists: The Myth of the Midwife-Witch," *Social History of Medicine* 3 (1990): 1–26.

6. See Susan R. Bordo, *The Flight to Objectivity: Essays on Cartesianism and Culture* (Albany: State University of New York Press, 1987), pp. 89, 101, 128; Carolyn Merchant, *The Death of Nature: Women, Ecology, and the Scientific Revolution* (New York: Harper and Row, 1980), pp. 132–40.

7. Cited by Wiesner, *Women and Gender,* p. 218. The stereotype of the witch is related on pp. 223–30. The early modern image of insatiable female sexuality and the assumption of melancholy in the temperament of women who believe they are witches is often traced to Robert Burton, *Anatomy of Melancholy,* 3 vols. (New York: Dutton, 1932). Burton's work was first published in 1621.

8. Barbara J. Shapiro, *Probability and Certainty in Seventeenth-Century England* (Princeton: Princeton University Press, 1983), pp. 194–226.

9. See n. 3 above; and Wiesner, *Women and Gender,* pp. 225–35.

10. See, e.g., an account of how witchcraft took on demonic qualities to the practitioners themselves as a direct result of Inquisitorial interrogations, in Carlo Ginsburg, *The Night Battles,* trans. John and Anne Tedeschi (Baltimore: Johns Hopkins University Press, 1983).

11. For example, the deluded view first put forth by Reginald Scot and Robert Burton in the early seventeenth century was taken up more strongly on evidential grounds by John Wagstaff in 1669 in *The Question of Witchcraft Debated,* cited by Shapiro, *Probability and Certainty,* pp. 217–18.

12. Merchant, *Death of Nature,* p. 141.

13. Shapiro, *Probability and Certainty,* pp. 195–97, 210–11; Wiesner, *Women and Gender,* pp. 223–25.

14. Shapiro, *Probability and Certainty,* pp. 209–11, 221–25.

15. Ibid., pp. 196, 205–10.

16. Ibid., pp. 198–204.

17. Ibid., pp. 212–20.

18. For an account of Locke's understanding of Newton and the importance of this view, see James A. Axtell, ed., *The Educational Writings of John Locke* (Cambridge: Cambridge University Press, 1968), pp. 84–87.

19. Isaac Newton, *Selections, Mathematical Principles of Natural Philosophy,* Florian Cajori revision of 1729 Andrew Motte trans. (Chicago: Henry Regnery, 1951), pp. 48–49.

20. John Henry, "Occult Qualities and the Experimental Philosophy: Active Principles in Pre-Newtonian Matter Theory," in *Essays in Early Modern Philosophy from Descartes and Hobbes to Newton and Leibniz,* vol. 7, *Seventeenth-Century Natural*

Scientists, ed. Vere Chappell (New York and London: Garland, 1992), pp. 1–47. The quote from Boyle is cited on p. 11.

21. Ibid., pp. 12–13.

22. Ibid., pp. 4–5, 30–34.

23. C. W. Lemmi, "Mythology and Alchemy in the Wisdom of the Ancients," in *The Classic Deities in Bacon: A Study in Mythological Symbolism* (Baltimore: Johns Hopkins University Press, 1933), pp. 46–49, 74–91, 196–213, reprinted in *Essential Articles for the study of Francis Bacon,* ed. Brian Vickers (Hamden, Conn: Archon Books, 1968), pp. 51–92.

24. Geoffrey Scarre, *Witchcraft and Magic in Sixteenth and Seventeenth Century Europe* (Atlantic Highlands, N.J.: Humanities Press, 1967), pp. 8–9.

25. Carolyn Merchant, *The Death of Nature: Women, Ecology and the Scientific Revolution* (New York: Harper and Row, 1980), pp. 136–38.

Notes to Chapter Fourteen

1. Maurice Cranston, *John Locke: A Biography* (London: Longmans Green, 1957), pp. 399–411.

2. Edward Neville Da Costa Andrade, *Sir Isaac Newton* (New York: Macmillan, 1954), pp. 92–99.

3. Michel Foucault, *The Order of Things* (New York: Vintage Books, 1973), pp. 166–80.

4. Ibid., pp. 182–85; Cranston, *Locke,* pp. 350–53.

5. Foucault, *Order,* pp. 180–89.

6. John Locke, *Two Treatises of Government,* ed. Peter Laslett (Cambridge: Cambridge University Press, 1991), Book II, Chap. v, Sec. 41, pp. 296–97. For a discussion of this issue that focuses on Locke's ideas about America and Indians, see Naomi Zack, "Locke and the Indians," in *The Social Power of Ideas,* ed. Yeager Hudson and Creighton Peden (Lewiston, N.Y.: Edwin Mellen Press, 1995), pp. 347–59.

7. Cited in Carolyn Merchant, *The Death of Nature: Women, Ecology and the Scientific Revolution* (New York: Harper and Row, 1980), p. 187.

8. See ibid., pp. 238, 246–48.

9. See n. 13 below.

10. Pashal Larkin, *Property in the Eighteenth Century* (London: Longmans Green, 1930), pp. 33–37; Merchant, *Death of Nature,* pp. 42–69.

11. F. D. Dow, *Radicalism in the English Revolution* (Oxford: Basil Blackwell, 1985), pp. 5, 33, 48, 79.

12. See above, chapter 6, n. 7.

13. Locke, *Two Treatises,* II, v, 46, p. 300.

14. Ibid., II, v, 47, pp. 300–301.

15. Ibid., II, v, 50, p. 302.

16. Merchant, *Death of Nature,* pp. 79–95.

17. Ibid., pp. 1–42.

18. See René Descartes, *Discourse on the Method of Rightly Conducting the Reason,* in *The Philosophical Works of Descartes,* ed. and trans. Elizabeth S. Haldane and G.R.T. Ross (Cambridge: Cambridge University Press, 1911; reprint 1962), vol. 1, Chap. V., pp. 109–10. John Cottingham points out that the mechanism of Cartesian animal bodies was in principle no different from the mechanism of Cartesian human bodies. See Cottingham, "'A Brute to the Brutes': Descartes' Treatment of Animals," in *Essays in Early Modern Philosophy from Descartes and Hobbes to Newton and Leibniz,* vol. 1, *René Descartes,* ed. Vere Chappell (New York and London: Garland, 1992), pp. 175–83.

19. See the discussion about Bacon's anti-feminist rhetoric about nature in chapter 1 above.

20. See Merchant, *Death of Nature,* pp. 253–75.

21. Ibid., pp. 236–42.

22. Ibid., p. 240.

23. Ibid., pp. 51–54.

24. Ibid., pp. 55–56.

25. For definitions of deep ecology as the ascription of intrinsic worth to natural beings, see Peter C. List, ed., *Radical Environmentalism* (Belmont, Calif.: Wadsworth, 1992), Chapter 1, articles by Arne Naess and Bill Devall and George Sessions, pp. 17–46.

26. On tectonic plate theory, see Keith Carlson, ed., *The Restless Earth,* Nobel Conference 24 (New York: Harper and Row, 1988).

27. See Jim Dodge, "Living by Life: Some Bioregional Theory and Practice," in List, *Radical Environmentalism,* pp. 108–16.

28. See Thomas S. Hill, Jr., "Ideals of Human Excellence and Preserving Natural Environments," in *Vice and Virtue in Everyday Life,* ed. Christina Sommers and Fred Sommers (New York: Harcourt Brace Jovanovich, 1989), pp. 293–310.

29. For discussion of the ways in which ecological damage can impact more negatively on dwellers in inner cities, see Cynthia Hamilton, "Women, Home, and Community: The Struggle in an Urban Environment"; and The Committee on Women, Population and the Environment, "Women, Population, and the Environment: Call for a New Approach," both in Allison M. Jagger, ed., *Living with Contradictions* (Boulder, Colo.: Westview Press, 1994), pp. 676–80, 694–97, respectively.

30. See Marshall D. Sahlins, "Notes on the Original Affluent Society," in *Man the Hunter,* ed. Richard B. Lee and Irven DeVore (Chicago: Aldene Publishing Co. 1968), pp. 85–89.

31. Paul Watson, "Tora! Tora! Tora!" in List, *Radical Environmentalism,* pp. 177–84.

Notes to the Afterword

1. Jean-Paul Sartre, *Anti-Semite and Jew,* trans. George J. Becker (New York: Schocken Books, 1965), p. 146.

Select Bibliography

Monographs and Anthologies

Andrade, Edward Neville Da Costa. *Sir Isaac Newton.* New York: Macmillan, 1954.

Ariès, Philippe. *Centuries of Childhood,* trans. Robert Baldick. New York: Alfred A. Knopf, 1962.

Aubrey, John. *Brief Lives,* ed. Richard Barber. Totowa, N.J.: Barnes and Noble, 1982.

Axtell, James A. *The Educational Writings of John Locke.* Cambridge: Cambridge University Press, 1968.

Ayers, Michael. *Locke, Volume I: Epistemology,* and *Volume II: Ontology.* New York: Routledge, 1991.

Bacon, Francis. *Essential Articles: Francis Bacon,* ed. Brian Vickers. Hamden: Archon Books, 1968.

Behn, Aphra. *Selected Writings of the Ingenious Mrs. Aphra Behn,* ed. Roberts Phelps. New York: Grove Press, 1950.

Blake, W. O. *The History of Slavery and the Slave Trade.* New York: Haskell House Publishers, 1858; reprint 1969.

Bordo, Susan R. *The Flight to Objectivity: Essays on Cartesianism and Culture.* Albany: State University of New York Press, 1987.

Chappell, Vere, ed. *Essays on Early Modern Philosophy from Descartes and Hobbes to Newton and Leibniz:* vol. 1, *René Descartes;* vol. 7, *Seventeenth-Century Natural Scientists.* New York: Garland, 1992.

Charleton, Walter. *Physiologia-Epicuro-Gassendo-Chartonia,* ed. Robert Heigh Kargon. New York: Johnson Reprint Corp., 1964.

Clark, Alice. *Working Life of Women in the Seventeenth Century,* ed. Miranda Chaytor and Jane Lewis. Boston and London: Routledge and Kegan Paul, 1982 (reprint from 1919).

Cranston, Maurice. *John Locke: A Biography.* London: Longmans Green, 1957.

Davies, K. G. *The Royal African Company.* London: Longmans Green, 1957.

Davis, David Brion. *Slavery and Human Progress.* Oxford: Oxford University Press, 1984.

De Morgan, Augustus. *Essays on the Life and Work of Newton,* ed. Philip E. B. Jourdain. Chicago and London: Open Court, 1914.

Descartes, René. *Meditations on First Philosophy,* in *Descartes' Philosophical Writings,* ed. Norman Kemp Smith. New York: Random House, 1958.

———. *The Philosophical Works of Descartes,* ed. and trans. Elizabeth Haldane and G.R.T. Ross. Cambridge: Cambridge University Press, 1911. Reprinted in 1962 in two vols.

Dow, F. D. *Radicalism in the English Revolution 1640–1660*, Oxford: Basil Blackwell, 1985.

Ehrenreich, Barbara, and Deidre English. *Witches, Midwives and Nurses: A History of Women Healers*. New York: Feminist Press, 1973.

Foucault, Michel. *The Order of Things*. New York: Vintage Books, 1973.

Frankfurt, Henry. *Demons, Dreamers, and Madmen*. New York: Bobbs-Merrill, 1970.

Funkenstein, Amos. *Theology and the Scientific Imagination*. Princeton: Princeton University Press, 1986.

Gassendi, Pierre. *The Selected Works of Pierre Gassendi,* ed. Craig B. Brush. New York: Johnson Reprint Corp., 1972.

Gaukroger, Stephen. *Descartes: An Intellectual Biography,* New York: Oxford University Press, 1995.

Ginsburg, Carlo. *The Night Battles,* trans. John and Anne Tedeschi. Baltimore: Johns Hopkins University Press, 1983.

Griffin, Martin I. J., Jr. *Latitudinarianism in the Seventeenth-Century Church of England,* annotated by Richard H. Popkin, ed. Lila Freedman. Leiden: E. J. Brill, 1992.

Hall, Marie Boas. *Robert Boyle on Natural Philosophy*. Bloomington: Indiana University Press, 1965.

Hobbes, Thomas. *The Citizen,* ed. Bernard Gert. New York: Anchor Books, 1992.

———. *The Leviathan,* ed. W. D. Pogson Smith. London: Oxford University Press, 1909. Reprinted in 1958.

Hole, Christina. *The English Housewife in the Seventeenth Century*. London: Chatto and Windus, 1953.

Jagger, Allison, ed. *Living with Contradictions*. Boulder, Colo.: Westview Press, 1994.

King, Lord [Peter], ed. *Life and Letters of John Locke*. New York: Lenox Hill, 1972.

Kuhn, Thomas S. *The Structure of Scientific Revolutions*. Chicago: University of Chicago Press, 1970.

Laquer, Thomas. *Making Sex: Body and Gender from the Greeks to Freud*. Cambridge: Harvard University Press, 1990.

Larkin, Pashal. *Property in the Eighteenth Century*. London: Longmans Green, l930.

Larner, Christina. *Enemies of God: The Witch Hunt in Scotland*. Baltimore: Johns Hopkins University Press. 1981.

———. *Witchcraft and Religion*. Oxford: Basil Blackwell, 1982.

List, Peter C., ed. *Radical Environmentalism.* Belmont, Calif: Wadsworth, 1992.

Lloyd, Genevieve. *The Man of Reason: "Male" and "Female" in Western Philosophy*. Minneapolis: University of Minnesota Press, 1984.

Locke, John. *An Essay Concerning Human Understanding,* ed. Peter H. Niddich. Oxford: Oxford University Press, l975.

———. *A Letter Concerning Toleration,* ed. James H. Tully. Indianapolis: Hackett, 1983.

———. *John Locke: Essays on the Law of Nature,* ed. W. von Leyden. Oxford: Oxford University Press, 1958.

———. *Two Treatises of Government,* ed. Peter Laslett. Cambridge: Cambridge University Press, 1991.

Lyndon, Marey Shanley, and Carole Pateman. *Feminist Interpretations and Political Theory*. University Park: Pennsylvania State University Press, 1991.

Lyons, Henry, F.R.S. *The Royal Society, 1660–1940*. New York: Greenwood Press, 1968.

MacInnes, C. M. *England and Slavery*. Bristol: Arrowsmith, 1934.

MacPherson, C. B. *The Political Theory of Possessive Individualism*. Oxford: Oxford University Press, 1962.

Maddison, R.E.W. *The Life of the Honorable Robert Boyle* (based on Thomas Birch's 1744 work). London: Taylor and Francis, 1969.

Mandelbaum, Maurice. *Philosophy, Science and Sense Perception*. Baltimore: Johns Hopkins University Press, 1964.

Mannix, D. P., and M. Cordley. *Black Cargoes: A History of the Atlantic Slave Trade.* New York: Viking, 1968.

Merchant, Carolyn. *The Death of Nature: Women, Ecology and the Scientific Revolution*. New York: Harper and Row, 1980.

Milton, John. *Areopagitica and Of Education,* ed. George H. Savine. New York: Appleton-Century-Croft, 1951.

Nicholson, Linda. *Gender and History*. New York: Columbia University Press, 1986.

———, ed. *Feminism/Postmodernism*. New York: Routledge, 1990.

Noonan, Harold W. *Personal Identity*. London: Routledge, l989.

Nozick, Robert. *Anarchy, State and Utopia*. New York: Basic Books, 1974.

Pepys, Samuel. *Everybody's Pepys: The Diary of Samuel Pepys 1660–1669*, ed. O. F. Morshead. London: G. Bell and Sons, 1948.

Popkin, Richard H. *The History of Skepticism from Erasmus to Spinoza*. Berkeley: University of California Press, 1979.

———, ed. *The Philosophy of the Sixteenth and Seventeenth Centuries*. New York: Collier, 1966.

Popkin, Richard H., and Charles B. Schmitt, eds. *Skepticism from the Renaissance to the Enlightenment*. Wiesbaden: Otto Harrossowitz, 1987.

Purdy, Laura M. *In Their Best Interest?* Ithaca and London: Cornell University Press, 1992.

Scharfstein, Ben-Ami. *The Philosophers, Their Lives and the Nature of Their Thought.* Oxford: Basil Blackwell, 1980.

Sedgwick, Eve Kosofsky. *Between Men: English Literature and Male Homosocial Desire.* New York: Columbia University Press, 1985.

Shapiro, Barbara J. *Probability and Certainty in Seventeenth-Century England.* Princeton: Princeton University Press, 1983.

Smith, Hilda L. *Reason's Disciples: Seventeenth-Century English Feminists.* Urbana: University of Illinois Press, 1982.

Sprat, Thomas. *History of the Royal Society,* ed. Jackson I. Cope and Harold Whitmore Jones. St. Louis: Washington University Press, 1958.

Stevens, James. *Francis Bacon and the Style of Science.* Chicago: University of Chicago Press, 1975.

Stimson, Dorothy. *Scientists and Amateurs: A History of the Royal Society.* New York: Henry Shuman, 1948.

Stone, Lawrence. *The Family, Sex and Marriage in England, 1500–1800.* New York: Harper and Row, 1979.

Tarcov, Nathan. *Locke's Education for Liberty.* Chicago: University of Chicago Press, 1984.

Thomas, Keith. *Religion and the Decline of Magic.* London: Weidenfeld and Nicolson, 1971.

Trevor-Roper, Hugh R. *The European Witch Craze of the Sixteenth and Seventeenth Centuries and Other Essays.* Harmondsworth: Penguin Books, 1969.

Wiesner, Merry E. *Women and Gender in Early Modern Europe.* Cambridge: Cambridge University Press, 1993.

Williams, Bernard. *Descartes: The Project of Pure Inquiry.* New York: Penguin, 1978.

Woolhouse, R. S. *The Empiricists.* Oxford: Oxford University Press, 1988.

Zack, Naomi. *Race and Mixed Race.* Philadelpha: Temple University Press, 1993.

Articles and Essays

Cohen, I. Bernard. "Hypotheses in Newton's Philosophy." In *Essays in Early Modern Philosophy from Descartes and Hobbes to Newton and Leibniz,* vol. 7, *Seventeenth-Century Natural Scientists,* ed. Vere Chappell, pp. 205–27. New York: Garland, 1992.

Cooper, Anthony Ashly, Third Earl of Shaftesbury. *Second Characters.* Reprinted in *Philosophies of Art and Beauty,* ed. Alfred Hofstadter and Richard Kuhns, pp. 239–76. Chicago: University of Chicago Press, 1964.

Day, John Patrick. "Self-Ownership." *Locke Newsletter* 20 (1989): 77–85.

Dubinin, N. P. "Race and Contemporary Genetics." In *Race, Science and Society,* ed. Leo Kuper, pp. 71–74. New York: Columbia University Press, 1965.

Hamilton, Cynthia. "Women, Home, and Community." In *Living with Contradictions,* ed. Allison M. Jagger. Boulder, Colo.: Westview Press, pp. 676–80.

Harley, David. "Historians As Demonologists: The Myth of the Midwife-Witch." *Social History of Medicine* 3 (1990): 1–26.

Henry, John. "Occult Qualities and the Experimental Philosophy: Active Principles in Pre-Newtonian Matter Theory." *Science History Publications Ltd.*, 1986. Reprinted in *Essays in Early Modern Philosophy from Descartes and Hobbes to Newton and Leibniz,* vol. 7, *Seventeenth Century Natural Scientists,* ed. Vere Chappell, pp. 1–48. New York: Garland, 1962.

Hill, Thomas S., Jr. "Ideals of Human Excellence and Preserving Natural Environments." In *Vice and Virtue in Everyday Life,* ed. Christina Sommers and Fred Sommers, pp. 327–44. New York: Harcourt Brace Jovanovitch, 1989.

Immerwahr, John, and Michael Burke. "Race and the Modern Philosophy Course." *Teaching Philosophy* 16:1 (March 1993): 21–34.

Law, Edmund. "A Defense of Mr. Locke's Opinion Concerning Personal Identity." In John Locke, *Works,* vol. 3. Scientia Verlag Aalen, 1963.

Lemmi, C. W. "Mythology and Alchemy in *The Wisdom of the Ancients.*" In *The Classic Deities in Bacon: A Study in Mythological Symbolism,* pp. 46–49, 74–91, 196–213. Baltimore: Johns Hopkins University Press, 1933. Reprinted in *Essential Articles for the study of Francis Bacon,* ed. Brian Vickers, pp. 51–92. Hamden, Conn.: Archon Books, 1968.

Mautner, Thomas. "Locke's Own." *Locke Newsletter* 22 (1991): 93–80.

McGuire, J. E. "Existence, Actuality and Necessity: Newton on Space and Time." In *Essays in Early Modern Philosophy from Descartes and Hobbes to Newton and Leibniz,* vol. 7, *Seventeenth-Century Natural Scientists,* ed. Vere Chappell, pp. 269–314. New York: Garland, 1992.

Milton, J. R. "John Locke and the *Fundamental Constitution* of Carolina." *Locke Newsletter* 21 (1990): 111–34.

Pateman, Carole. "'God Hath Ordained to Man a Helper': Hobbes, Patriarchy and Conjugal Right." In *Feminist Interpretations and Political Theory,* ed. Mary Lyndon Shanley and Carole Pateman, pp. 53–73. University Park: University of Pennsylvania Press, 1991.

Popkin, R. H. "The Skeptical Crisis and the Rise of Modern Philosophy," Parts I, II, and III. *Review of Metaphysics* 8 (1953–54): 131–51, 307–34, 499–510.

Sahlins, Marshall D. "Notes on the Original Affluent Society." In *Man the Hunter,* ed. Richard B. Lee and Irven DeVore, pp. 85–89. Chicago: Aldine Publishing Co., 1968.

Scharfstein, Ben-Ami. "Descartes' Dreams." *Philosophical Forum* 1 N.S., 3 (1969): 302–21.

van Fraassen, Bas. "To Save the Phenomena." *Journal of Philosophy* 18 (Oct. 1976): 623–32.

Winkler, Kenneth P. "Locke on Personal Identity." *Journal of the History of Philosophy* 29:2 (April 1991): 214–17.

Zack, Naomi. "Locke's Identity Meaning of Ownership." *Locke Newsletter* 23 (1992): 105–14.

———. "Mixed Black and White Race and Public Policy." *Hypatia* 10:1 (February 1995): 120–32.

———. "Race and Philosophic Meaning." *American Philosophical Association Newsletter on Philosophy and the Black Experience* 94:1 (Fall 1994): 11–18.

———. "Locke and the Indians." In *The Social Power of Ideas,* ed. Yeager Hudson and Creighton Peden, pp. 347–59. Lewiston, N.Y.: Edwin Mellen Press, 1995.

Index